Cloud Computing
with
e-Science Applications

Cloud Computing
with
e-Science Applications

EDITED BY
OLIVIER TERZO
ISMB, TURIN, ITALY

LORENZO MOSSUCCA
ISMB, TURIN, ITALY

CRC Press
Taylor & Francis Group
Boca Raton London New York

CRC Press is an imprint of the
Taylor & Francis Group, an **informa** business

CRC Press
Taylor & Francis Group
6000 Broken Sound Parkway NW, Suite 300
Boca Raton, FL 33487-2742

First issued in paperback 2020

© 2015 by Taylor & Francis Group, LLC
CRC Press is an imprint of Taylor & Francis Group, an Informa business

No claim to original U.S. Government works

ISBN-13: 978-1-4665-9115-8 (hbk)
ISBN-13: 978-0-367-73853-2 (pbk)

Library of Congress Cataloging-in-Publication Data

Cloud computing with e-science applications / editors, Olivier Terzo, Lorenzo Mossucca.
      pages cm
   Includes bibliographical references and index.
   ISBN 978-1-4665-9115-8 (hardback)
      1. Research--Data processing. 2. Science--Data processing. 3. Cloud computing. I. Terzo, Olivier. II. Mossucca, Lorenzo.

   Q183.9.C58 2015
   502.85'46782--dc23   2015004557

Visit the Taylor & Francis Web site at
http://www.taylorandfrancis.com

and the CRC Press Web site at
http://www.crcpress.com

Contents

Preface

The interest in cloud computing in both industry and research domains is continuously increasing to address new challenges of data management, computational requirements, and flexibility based on needs of scientific communities, such as custom software environments and architectures. It provides cloud platforms in which users interact with applications remotely over the Internet, bringing several advantages for sharing data, for both applications and end users. Cloud computing provides everything: computing power, computing infrastructure, applications, business processes, storage, and interfaces, and can provide services wherever and whenever needed.

Cloud computing provides four essential characteristics: elasticity; scalability; dynamic provisioning of applications, storage, and resources; and billing and metering of service usage in a pay-as-you-go model. This flexibility of management and resource optimization is also what attracts the main scientific communities to migrate their applications to the cloud.

Scientific applications often are based on access to large legacy data sets and application software libraries. Usually, these applications run in dedicated high performance computing (HPC) centers with a low-latency interconnection. The main cloud features, such as customized environments, flexibility, and elasticity, could provide significant benefits.

Since every day the amount of data is exploding, this book describes how cloud computing technology can help such scientific communities as bioinformatics, earth science, and many others, especially in scientific domains where large data sets are produced. Data in more scenarios must be captured, communicated, aggregated, stored, and analyzed, which opens new challenges in terms of tool development for data and resource management, such as a federation of cloud infrastructures and automatic discovery of services.

Cloud computing has become a platform for scalable services and delivery in the field of services computing. Our intention is to put the emphasis on scientific applications using solutions based on cloud computing models—public, private, and hybrid—with innovative methods, including data capture, storage, sharing, analysis, and visualization for scientific algorithms needed for a variety of fields. The intended audience includes those who work in industry, students, professors, and researchers from information technology, computer science, computer engineering, bioinformatics, science, and business fields.

Actually, applications migration in the cloud is common, but a deep analysis is important to focus on such main aspects as security, privacy, flexibility, resource optimization, and energy consumption.

This book has 12 chapters; the first two are on exposing a proposal strategy to move applications in the cloud. The other chapters are a selection of some applications used on the cloud, including simulations on public transport, biological analysis, geographic information system (GIS) applications, and more. Various chapters come from research centers, universities, and industries worldwide: Singapore, Australia, China, Hong Kong, India, Brazil, Colombia, the Netherlands, Germany, the United Kingdom, Hungary, Spain, and Ireland. All contributions are significant; most of the research leading to results has received funding from European and regional projects.

After a brief overview of cloud models provided by the National Institute of Standards and Technology (NIST), Chapter 1 presents several criteria to meet user requirements in e-science fields. The cloud computing model has many possible combinations; the public cloud offers an alternative to avoid the up-front cost of buying dedicated hardware. Preliminary analysis of user requirements using specific criteria will be a strong help for users for the development of e-science services in the cloud.

Chapter 2 discusses the challenges that are imposed by big data on scientific data infrastructures. A definition of big data is shown, presenting the main application fields and its characteristics: volume, velocity, variety, value, and veracity. After identifying research infrastructure requirements, an e-science data infrastructure is introduced using cloud technology to answer future big data requirements. This chapter focuses on security and trust issues in handling data and summarizes specific requirements to access data. Requirements are defined by the European Research Area (ERA) for infrastructure facility, data-processing and management functionalities, access control, and security.

One of the important aspects in the cloud is certainly security due to the use of personal and sensitive information, especially derived mainly by social network and health information. Chapter 3 presents a set of important vulnerability issues, such as data theft or loss, privacy issues, infected applications, threats in virtualization, and cross-virtual machine attack. Many techniques are used to protect against cloud service providers, such as homomorphic encryption, access control using attributes based on encryption, and data auditing through provable data possession and proofs of irretrievability. The chapter underlines points that are still open, such as security in the mobile cloud, distributed data auditing for clouds, and secure multiparty computation on the cloud.

Many e-science applications can be modeled as workflow applications, defined as a set of tasks dependent on each other. Cloud technology and platforms are a possible solution for hosting these applications. Chapter 4 discusses implementation aspects for execution of workflows in clouds. The proposal architecture is composed of two layers: platform and application. The first one, described as scientific workflow, enables operations such as dynamic resource provisioning, automatic scheduling of applications, fault tolerance, security, and privacy in data access. The second one defines data analytic applications enabling simulation of the public transport system of Singapore and the effect of unusual events in its network. This application provides evaluation of the effect of incidents in the flow of passengers in that country.

Chapter 5 presents the main aspects for the cloud characterization and design on a large amount of data and intensive computational context. A new version of migration methodology derived by Laszewski and Nauduri algorithms is introduced. Then, it discusses the realization of a free cloud data migration tool for the migration of the database in the cloud and the refactoring of the application architecture. This tool provides two main functionalities: storage for cloud data and cloud data services. It allows supporting target adapters for several data stores and services such as Amazon RDS, MongoDB, Mysql, and so on. The chapter concludes with an evaluation of migration of the SimTech Scientific Workflow Management System to Amazon Web Services. Results of this research have mainly received funding from the project 4CaaSt (from the European Union's Seventh Framework Programme) and from the German Research Foundation within the Cluster of Excellence in Simulation Technology at the University of Stuttgart.

Chapter 6 presents a proposal developed under the e-Clouds project for a scientific software-as-a-service (SaaS) marketplace based on the utilization of the resource provided by a public infrastructure-as-a-service (IaaS) infrastructure, allowing various users to access on-demand applications. It automatically manages the complexity of configuration required by public IaaS providers by delivering a ready environment for using scientific applications, focusing on the different patterns applied for cloud resources while hiding the complexity for the end user. Data used for testing architecture comes from the Alexander von Humboldt Institute for Biological Resources.

A systematic way of building a web-based geographic information system is presented in Chapter 7. Key elements of this methodology are a database management system (DBMS), base maps, a web server with related storage, and a secure Internet connection. The application is designed for analyzing the main causes of road accidents and road state and quality in specific regions. Local organizations can use this information to organize preventive measures for reducing road accidents. Services and applications have been deployed in the main public cloud platforms: Microsoft Windows Azure platform and Amazon Web Service. This work has been partly funded by the Horizon Fund for Universities of the Scottish Funding Council.

The physical and psychological pressures on people are increasing constantly, which raises the potential risks of many chronic diseases, such as high blood pressure, diabetes, and coronary disease. Cloud computing has been applied to several real-life scenarios, and with the rapid progress in its capacity, more and more applications are provided as a service mode (e.g., security as a service, testing as a service, database as a service, and even everything as a service). Health care service is one such important application field. In Chapter 8, a ubiquitous health care system, named HCloud, is described; it is a smart information system that can provide people with some basic health monitoring and physiological index analysis services and provide an early warning mechanism for chronic diseases. This platform is composed of physiological data storage, computing, data mining, and several features. In addition, an online analysis scheme combined with the MapReduce parallel framework is designed to improve the platform's capabilities. The MapReduce paradigm has features of code simplicity, data splitting, and automatic parallelization compared with other distributed parallel systems, improving efficiency of physiological data processing and achieving increased linear speed.

With the explosive growth in the use of information and communication technology, applications that involve deep analytics in a big data scenario need to be shifted to a scalable context. A noticeable effort has been made to move the data management systems into MapReduce parallel processing environments. Chapter 9 presents RPig, an integrated framework with R and Pig for scalable machine learning and advanced statistical functionalities, which makes it feasible to use high-level languages to develop analytic jobs easily in concise programming. RPig benefits from the deep statistical analysis capability of R and parallel data-processing capability of Pig.

Parameter sweep applications are frequent in scientific simulations and in other types of scientific applications. Cloud computing infrastructures are suitable for these kinds of applications due to their elasticity and ease of scaling up on demand. They run the same application with a very large number of parameters; hence, execution time could take very long on a single computing resource. Chapter 10 presents the AutoDock program for modeling intermolecular interactions. It provides a suite of automated docking tools designed to predict how small molecules, such as substrates or drug candidates, bind to a receptor of known three-dimensional (3D) structure. The proposed solutions are tailored to a specific grid or cloud environment. Three different parameter sweep workflows were developed and supported by the European Commission's Seventh Framework Programme under projects SCI-BUS and ER-Flow.

There are also disadvantages to using applications in the cloud, such as usability issues in IaaS clouds, limited language support in platform-as-a-service clouds, and lack of specialized services in SaaS clouds. For resolving known issues, Chapter 11 proposes the development of research clouds for high-performance computing as a service (HPCaaS) to enable researchers to take on the role of cloud service developer. It consists of a new cloud model, HPCaaS, which automatically configures cloud resources for HPC. An SaaS cloud framework to support genomic and medical research is presented that allows simplifying the procedures undertaken by service providers, particularly during service deployment. By identifying and automating common procedures, the time and knowledge required to develop cloud services is minimized. This framework, called Uncino, incorporates methodologies used by current e-science and research clouds to simplify the development of SaaS applications; the prototype is compatible with Amazon EC2, demonstrating how cloud platforms can simplify genomic drug discovery via access to cheap, on-demand HPC facilities.

e-Science applications such as the ones found in Smart Cities, e-Health, or Ambient Intelligence require constant high computational demands to capture, process, aggregate, and analyze data. Research is focusing on the energy consumption of the sensor deployments that support this kind of application. Chapter 12 proposes global energy optimization policies that start from the architecture design of the system, with a deeper focus on data center infrastructures (scheduling and resource allocation) and take into account the energy relationship between the different abstraction layers, leveraging the benefits of heterogeneity and application awareness. Data centers are not the only computing resources involving energy inefficiency; distributed computing devices and wireless communication layers also are included. To provide adequate energy management, the system is tightly coupled with an energy analysis and an optimization system.

Acknowledgments

We would like to express our gratitude to all the professors and researchers who contributed to this, our first, book and to all those who provided support, talked things over, or read, wrote, and offered comments.

We thank all authors and their organizations that allowed sharing relevant studies of scientific applications in cloud computing and thank advisory board members Fatos Xhafa, Hamid R. Arabnia, Vassil Alexandrov, Pavan Balaji, Harold Enrique Castro Barrera, Rajdeep Bhowmik, Michael Gerhards, Khalid Mohiuddin, Philippe Navaux, Suraj Pandey, and Ioan Raicu for providing important comments to improve the book.

We wish to thank our research center, Istituto Superiore Mario Boella, which allowed us to become researchers in the cloud computing field, especially our director, Dr. Giovanni Colombo; our deputy director of the research area, Dr. Paolo Mulassano; our colleagues from research unit IS4AC (Infrastructure and System for Advanced Computing): Pietro Ruiu, Giuseppe Caragnano, Klodiana Goga, and Antonio Attanasio, who supported us in the reviews.

A special thanks to our publisher, Nora Konopka, for allowing this book to be published and all persons from Taylor & Francis Group who provided help and support at each step of the writing.

We want to offer a sincere thank you to all the readers and all persons who will promote this book.

Olivier Terzo and Lorenzo Mossucca

About the Editors

Olivier Terzo is a senior researcher at Istituto Superiore Mario Boella (ISMB). After receiving a university degree in electrical engineering technology and industrial informatics at the University Institute of Nancy (France), he received an MSc degree in computer engineering and a PhD in electronic engineering and communications from the Polytechnic of Turin (Italy).

From 2004 to 2009, Terzo was a researcher in the e-security laboratory, mainly with a focus on P2P (peer-to-peer) protocols, encryption on embedded devices, security of routing protocols, and activities on grid computing infrastructures. From 2010 to 2013, he was the head of the Research Unit Infrastructures and Systems for Advanced Computing (IS4AC) at ISMB.

Since 2013, Terzo has been the head of the Research Area: Advanced Computing and Electromagnetics (ACE), dedicated to the study and implementation of computing infrastructure based on virtual grid and cloud computing and to the realization of theoretical and experimental activities of antennas, electromagnetic compatibility, and applied electromagnetics.

His research interest focuses on hybrid private and public cloud distributed infrastructure, grid, and virtual grid; mainly, his activities involve application integration in cloud environments. He has published about 60 papers in conference proceedings and journals, and as book chapters.

Terzo is also involved in workshop organization and the program committee of the CISIS conference; is an associate editor of the *International Journal of Grid and Utility Computing* (IJGUC); International Program Committee (IPC) member of the International Workshop on Scalable Optimisation in Intelligent Networking; and peer reviewer in International Conference on Networking and Services (ICNS) and International Conference on Complex Intelligent and Software Intensive Systems (CISIS) conferences.

Dr. Lorenzo Mossucca studied computer engineering at the Polytechnic of Turin. From 2007, he has worked as a researcher at the ISMB in IS4AC.

His research interests include studies of distributed databases, distributed infrastructures, and grid and cloud computing. For the past few years, he has focused his research on migration of scientific applications to the cloud, particularly in the bioinformatics and earth sciences fields.

He has published about 30 papers in conference proceedings, journals, and posters and as chapters.

He is part of the Technical Program Committee and is a reviewer for many international conferences, including the International Conference on Complex, Intelligent, and Software Intensive Systems, International Conference on Networking and Services, and Institute of Electrical and Electronics Engineers (IEEE) International Symposium on Parallel and Distributed Processing with Applications and journals such as *IEEE Transactions on Services Computing, International Journal of Services Computing, International Journal of High Performance Computing and Networking,* and *International Journal of Cloud Computing.*

List of Contributors

Muhammad Akmal
Pisys Limited
Aberdeen, United Kingdom

Ian Allison
Robert Gordon University
Aberdeen, United Kingdom

Vasilios Andrikopoulos
Institute of Architecture of
    Application Systems (IAAS)
University of Stuttgart
Stuttgart, Germany

Patricia Arroba
Electronic Engineering Department
Universidad Politécnica de Madrid
Madrid, Spain

José Luis Ayala Rodrigo
Departamento de Arquitectura
    de Computadores y Automática
    (DACYA)
Universidad Complutense de Madrid
Madrid, Spain

Ákos Balaskó
Institute for Computer Science and
    Control of the Hungarian Academy
    of Sciences (MTA SZTAKI)
Budapest, Hungary

Péter Borsody
University of Westminster
London, United Kingdom

Rajkumar Buyya
Cloud Computing and Distributed
    Systems Lab
Department of Computing
    and Information Systems
University of Melbourne
Melbourne, Australia

Yunpeng Cai
Shenzhen Institutes of Advanced
    Technology
Chinese Academy of Sciences
Beijing, China

Rodrigo N. Calheiros
Cloud Computing and Distributed
    Systems Lab
Department of Computing and
    Information Systems
University of Melbourne
Melbourne, Australia

Harold Castro
Communications and Information
    Technology Group (COMIT)
Department of Systems and
    Computing Engineering
Universidad de los Andes
Bogotá, Colombia

Philip Church
School of IT
Deakin University
Highton, Australia

Alexandre da Silva Carissimi
Federal University of Rio Grande
    do Sul
Porto Alegre, Brazil

Cees de Laat
System and Network Engineering
    Group
University of Amsterdam,
    Netherlands

Yuri Demchenko
System and Network Engineering
    Group
University of Amsterdam
Amsterdam, Netherlands

Xiaomao Fan
Shenzhen Institutes of Advanced
    Technology
Chinese Academy of Sciences
Beijing, China

Zoltán Farkas
Institute for Computer Science and
    Control of the Hungarian Academy
    of Sciences (MTA SZTAKI)
Budapest, Hungary

José Manuel Moya Fernandez
Electronic Engineering Department
Universidad Politécnica de Madrid
Madrid, Spain

Horacio Gonzalez-Velez
National College of Ireland
Dublin, Ireland

Andrzej Goscinski
School of IT
Deakin University
Geelong, Australia

Paola Grosso
System and Network Engineering
    Group
University of Amsterdam,
    Netherlands

Ákos Hajnal
Institute for Computer Science and
    Control of the Hungarian Academy
    of Sciences (MTA SZTAKI)
Budapest, Hungary

Sidath B. Handurukande
Network Management Lab
Ericsson
Athlone, Ireland

Chenguang He
Shenzhen Institutes of Advanced
    Technology
Chinese Academy of Sciences
Beijing, China

Katzalin Olcoz Herrero
Departamento de Arquitectura
    de Computadores y Automática
    (DACYA)
Universidad Complutense de Madrid
Madrid, Spain

Xucan Huang
Shenzhen Institutes of Advanced
    Technology
Chinese Academy of Sciences
Beijing, China

Terence Hung
Institute of High Performance
    Computing
A*STAR Institute
Singapore

Péter Kacsuk
Institute for Computer Science and
    Control of the Hungarian Academy
    of Sciences (MTA SZTAKI)
Budapest, Hungary

Dimka Karastoyanova
Institute of Architecture
  of Application Systems (IAAS)
University of Stuttgart
Stuttgart, Germany

Krisztián Karóczkai
Institute for Computer Science and
  Control of the Hungarian Academy
  of Sciences (MTA SZTAKI)
Budapest, Hungary

Henry Kasim
Institute of High Performance
  Computing
A*STAR Institute
Singapore

Tamás Kiss
University of Westminster
London, United Kingdom

Gary Lee
Institute of High Performance
  Computing
A*STAR Institute
Singapore

Xiaorong Li
Institute of High Performance
  Computing
A*STAR Institute
Singapore

Ye Li
Shenzhen Institutes of Advanced
  Technology
Chinese Academy of Sciences
Beijing, China

Sifei Lu
Institute of High Performance
  Computing
A*STAR Institute
Singapore

Peter Membrey
Hong Kong Polytechnic University
Hong Kong

**Philippe Olivier Alexandre
Navaux**
Federal University of Rio Grande
  do Sul
Porto Alegre, Brazil

Canh Ngo
System and Network Engineering
  Group
University of Amsterdam
Amsterdam, Netherlands

Tuan Ngo
Department of Infrastructure
  Engineering
University of Melbourne
Melbourne, Australia

Henry Novianus Palit
Petra Christian University
Surabaya, Indonesia

Eduardo Roloff
Federal University of Rio Grande
  do Sul
Porto Alegre, Brazil

Sushmita Ruj
R. C. Bose Center for Cryptology
  and Security
Indian Statistical Institute
Kolkata, India

Rajat Saxena
School of Computer Science
  and Engineering
Indian Institute of Technology
Indore, India

Steve Strauch
Institute of Architecture
    of Application Systems (IAAS)
University of Stuttgart
Stuttgart, Germany

David Susa
Communications and Information
    Technology Group (COMIT)
Department of Systems
    and Computing Engineering
Universidad de los Andes
Bogotá, Colombia

Zahir Tari
School of Computer Science and IT
RMIT University
Melbourne, Australia

Mario Villamizar
Communications and Information
    Technology Group (COMIT)
Department of Systems
    and Computing Engineering
Universidad de los Andes
Bogotá, Colombia

Karolina Vukojevic-Haupt
Institute of Architecture
    of Application Systems (IAAS)
University of Stuttgart
Stuttgart, Germany

Long Wang
Institute of High Performance
    Computing
A*STAR Institute
Singapore

MingXue Wang
Network Management Lab
Ericsson
Athlone, Ireland

Adam Wong
George Washington University
Ashburn, Virginia, USA

Marina Zapater
CEI Campus Moncloa UCM-UPM
Madrid, Spain

1

Evaluation Criteria to Run Scientific Applications in the Cloud

**Eduardo Roloff, Alexandre da Silva Carissimi,
and Philippe Olivier Alexandre Navaux**

CONTENTS

Summary

In this chapter, we will present a brief explanation of the services and implementation of models of cloud computing in order to promote a discussion of the strong and weak points of each. Our aim is to select the best combination of the models as a platform for executing e-science applications.

Additionally, the evaluation criteria will be introduced so as to guide the user in making the correct choice from the available options. After that, the main public cloud providers, and their chief characteristics, are discussed.

One of the most important aspects of choosing a public cloud provider is the cost of its services, but its performance also needs to be taken into account. For this reason, we have introduced the cost efficiency evaluation to support the user in assessing both price and performance when choosing a provider. Finally, we provide a concrete example of applying the cost efficiency evaluation using a real-life situation and including our conclusions.

1.1 Introduction

To create a service to execute scientific applications in the cloud, the user needs to choose an adequate cloud environment [1, 2]. The cloud computing model has several possible combinations between the service and implementation models, and these combinations need to be analyzed. The public cloud providers offer an alternative to avoid the up-front costs of buying machines, but it is necessary to evaluate them using certain criteria to verify if they meet the needs of the users. This chapter provides a discussion about these aspects to help the user in the process of building an e-Science service in the cloud.

1.2 Cloud Service Models

According to the National Institute of Standards and Technology (NIST) definition [3], there are three cloud service models, represented in Figure 1.1. They present several characteristics that need to be known by the user. All three models have strong and weak points that influence the adequacy for use to create an e-Science service.

The characteristics of the service models are presented and discussed in this section.

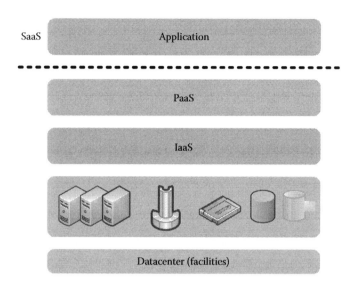

FIGURE 1.1
Service models.

1.2.1 Software as a Service

The software-as-a-service (SaaS) model is commonly used to deliver e-science services to users. This kind of portal is used to run standard scientific applications, and no customization is allowed. Normally, a provider ports an application to its cloud environment and then provides access for the users to use the applications on a regular pay-per-use model. The user of this model is the end user, such as a biologist, and there is usually no need to modify the application.

One example of a provider porting a scientific application and then providing the service to the community is the Azure BLAST [2] project. In this project, Microsoft ports the Basic Local Alignment Search Tool (BLAST) of the National Center for Biotechnology Information (NCBI) to Windows Azure. BLAST is a suite of programs used by bioinformatics laboratories to analyze genomics data. Another case of this use are the Cyclone Applications, which consist of twenty applications offered as a service by Silicon Graphics Incorporated (SGI). SGI provides a broad range of applications that cover several research topics, but there is no possibility to customize and adapt them.

The big problem with SaaS as the environment to build e-science services is the absence of the ability for customization. Research groups are constantly improving their applications, adding new features, or improving their performance, and they need an environment to deliver the modifications. In addition, there are several applications that are used for only a few research groups, and this kind of application does not attract the interest of the cloud providers to port them. In this case, this model can be used to deliver an e-science service but not as an environment to build it.

1.2.2 Platform as a Service

The platform-as-a-service (PaaS) model presents more flexibility than the SaaS model. Using this model, it is possible to develop a new, fully customized application and then execute it in the provider's cloud environment. It is also possible to modify an existing application to be compatible with the provider's model of execution; in the majority of cases, this is a realistic scenario for scientific applications [4]. The majority of the services provided in this model consist of an environment to execute web-based applications. This kind of application processes a large number of simultaneous requests from different users. The regular architecture of these applications is composed of a web page, which interacts with the user; a processing layer, which implements the business model; and a database, used for data persistence. Each user request is treated uniquely in the system and has no relationship with other requests. Due to this, it is impossible to create a system to perform distributed computing. However, the processing layer of this model can be used if the service does not have a huge demand for processing power.

In the PaaS model, the provider defines the programming languages and the operating system that can be used; this is a limitation for general-purpose scientific application development.

1.2.3 Infrastructure as a Service

The infrastructure-as-a-service (IaaS) model is the most flexible service model of cloud computing. The model delivers raw computational resources to the user, normally in the form of virtual machines (VMs). It is possible to choose the size of the VM, defining the number of cores and the amount of memory. The user can even choose the operating system and install any desired software in the VM. The user can allocate any desired quantity of VMs and build a complete parallel system. With this flexibility, it is possible to use IaaS for applications that need a large amount of resources by the configuration of a cluster in the cloud.

1.3 Cloud Implementation Models

The service models, presented in the previous section, can be delivered using four different implementation models: private cloud, community cloud, public cloud, and hybrid cloud. Each one has strong and weak points. The four models can be used to build an e-science service, and they are analyzed to present their main characteristics to help the user decide which one to choose.

1.3.1 Private Cloud

A private cloud is basically the same as owning and maintaining a traditional cluster, where the user has total control over the infrastructure and can configure the machines according to need. One big issue in a private scenario is the absence of instant scalability, as the capacity of execution is limited to the physical hardware available. Moreover, the user needs to have access to facilities to maintain the machines and is responsible for the energy consumption of the system. Another disadvantage is the hardware maintenance; for example, if a machine has physical problems, the user is responsible for fixing or replacing it. A case for which the private cloud is recommended is if the application uses confidential or restricted data; in this scenario, the access control to the data is guaranteed by the user's policies. The weakness of this model is the absence of elasticity and the need for up-front costs. Building a private cloud for scientific applications can be considered the same as buying a cluster system.

1.3.2 Community Cloud

In a community cloud, the users are members of one organization, and this organization has a set of resources that are connected to resources in other organizations. A user from one of the organizations can use the resources of all other organizations. The advantage of this model is the provision of access to a large set of resources without charging because the remote resources belong to other organizations that form the community and not to a provider. In other words, the pay-per-use model may not be applicable to this type of cloud. One disadvantage of the model is the limited number of resources; they are limited to the number of machines that are part of the community cloud. The interconnection between all the members constitutes a bottleneck for the application's execution. If the application needs more machines than are available in single site (a single member), the machines need to be allocated among two or more members.

All the community members need to use the same cloud platform; this demands an effort to configure all the machines, and it is necessary to have personnel to maintain the machines. The community model is recommended for research groups that are geographically distributed and want to share the resources among them.

1.3.3 Public Cloud

In a public cloud, the infrastructure is provided by a company, the provider. The advantage in this case is the access to an unlimited number of computational resources, where the user can allocate and deallocate them according to demand. The pay-per-use billing model is also an advantage because the user has to spend money only while using the resources. Access to up-to-date hardware without the up-front costs and the absence of maintenance costs complete the list of advantages of the public model. The main disadvantages relate to data privacy because, in this model, the underlying hardware belongs to a provider, and all the maintenance procedures are made by the provider's personnel. The data privacy issue can be addressed by a contract regarding data access, but for certain types of users, such as banks, this is insufficient. The user has access to virtualized hardware controlled by a hypervisor and does not have control over the underlying resources, such as physical machines and network infrastructure. In this model, the user has access only to a virtual environment; sometimes, this can be insufficient. Certain applications need specific hardware configurations to reach acceptable performance levels, and these configurations cannot be made in a public cloud environment. The recommended scenario to use this model is if the user needs to execute an application during a limited time period, and this is an advantage for an e-science service. Moreover, in case of an application executing only a few hours a day, the user can allocate the machines, execute the application, and deallocate the machines; the user just needs to pay for the time used. Even if the application will run during almost the entire day, without a predefined end date, it is necessary to determine the cost-benefit ratio of using a public cloud instead of buying physical machines.

1.3.4 Hybrid Cloud

A hybrid cloud can be used to extend the computational power available on a user-owned infrastructure with a connection to an external provider. This model is recommended if the user needs to increase the capacity of the user's infrastructure without the acquisition of new hardware. The main advantage of it is the instant access to computational power without up-front costs. In certain scenarios, it is possible to configure the system to allocate resources in the cloud automatically, with the system allocating and deallocating machines according to demand. This model is applicable if the user already has a set of machines and needs to increase them temporarily, for example, for a specific project.

The weakness of this model is related to data transfer because the local cloud is connected to the public cloud through a remote connection, normally an Internet connection; in this case, the bandwidth is limited by this connection. In an application that has a large amount of communication, the connection between the user and provider will be the bottleneck and can affect the overall performance. Another important issue is the cloud platform used by the cloud provider. It is necessary that the user's system use the same platform, or at least a compatible one. This means that the user needs to reconfigure all the local machines to follow the cloud model. The concerns about data confidentiality are the same as in the public model.

TABLE 1.1

Comparison of Implementation Models

	Advantage	Drawback
Private	Privacy	Scalability
Community	Shared cost	Scalability
Public	Scalability	Privacy
Hybrid	Scalability	Interconnection

1.3.5 Summary of the Implementation Models

Summarizing the characteristics presented in this section, we can conclude that all deployment models can be used to create high-performance computing (HPC) environments in the cloud. The appropriate model depends on the needs of the user and the user's available funds. All the models have advantages and disadvantages, and it is clear that there is no ideal model for all the usage scenarios. Table 1.1 summarizes the main advantage and disadvantage of each cloud implementation model.

1.4 Considerations about Public Providers

The private and community models are well known by users due to their similarity to clusters and grids. The hybrid and public models are really new paradigms of computing. As the hybrid model is a combination of local machines and a public provider, we can conclude that the new paradigm is the public cloud. In the rest of this chapter, we perform an analysis of the public cloud model.

When choosing a public cloud provider, the user needs to consider relevant aspects of his service. Some of these concerns are explained here. However, the user needs to perform an analysis of the necessary service level for his service.

1.4.1 Data Confidentiality

Data confidentiality is one of the main concerns regarding public cloud providers. In addition, relevant aspects about data manipulation need to be considered:

- **Segregation:** The provider needs to guarantee data segregation between clients because most of them use shared resources. It is necessary to ensure that the user's data can only be accessed by authorized users.

- **Recovery and backup procedures**: The user needs to evaluate the backup procedures of the provider. All the backup tapes need to be encrypted to maintain data confidentiality. Also, the recovery procedures need to be well documented and tested on a regular basis.

- **Transfer**: It is necessary that the provider implements secure data transfer between the user and provider. Also, standard transfer mechanisms should be provided to the user to implement in the user's applications.

1.4.2 Administrative Concerns

Most of the administrative concerns need to be covered in the contract between the user and the provider and need to be well described. It is necessary to choose a provider with an adequate service-level agreement (SLA). Normally, the SLA is standard for all the users, but in the case of special needs, it is possible to negotiate with the provider. Also, the penalties if the SLA is not correctly delivered can be added to the contract. In most cases, changes in the standard SLA incur extra costs.

The provider must deliver a monitoring mechanism to the user to verify system health and the capacity of its allocated resources. Reporting tools are necessary to evaluate all the quality and usage levels.

The billing method is another important point of attention; it is necessary to know how the provider charges the user. In many cases, the smallest unit to charge a VM is 1 hour, even if it was used just for 5 minutes. Some providers present costs related to data transfer to outside the cloud. The storage price is another concern; some providers have free storage, up to a certain amount, and others charge in different manners. All the costs incurred in the operation need to be known by the user and controlled by the provider.

The provider's business continuity is also an aspect to take into account. This is an administrative and technical concern. In the case of the provider's end of the business, it is necessary that the user have guaranteed access to his or her own data. Also, the user needs the capability to move data to another provider without much effort; this is an important interoperability aspect.

1.4.3 Performance

A typical public cloud computing environment is a hosted service available on the Internet. The user needs to be continuously connected to the cloud provider with the agreed speed, both for data transfer from and to the provider and for regular access to the provider's cloud manager. The Internet connection speed and availability are an issue even for performance and reliability with a cloud computing service.

The major issues regarding performance in cloud computing is the virtualization and network interconnection. If the hypervisor does not have good resource management, it is possible that the physical resources are under- or overused. In this case, a user can allocate a VM instance of a certain size and when the VM is moved to other resources of the provider's infrastructure, the processing performance decreases or increases. Also, the network interconnection of the VM is a concern; as the network resources are pooled among all the users, the network performance is not guaranteed. This is an important topic for applications that use a large number of instances.

1.5 Evaluation Criteria

To provide a comprehensive evaluation of cloud computing as an environment for e-science services, for both technical and economic criteria, it is necessary to evaluate three aspects.

- **Deployment**: This aspect is related to the deployment capability of providers to build e-science environments in the cloud and the capability to execute the workload.
- **Performance**: This is the performance evaluation of the cloud compared to a traditional machine.
- **Economic**: The economic evaluation is performed to determine if it is better to use a cloud or to buy regular machines.

The deployment capability of cloud computing relates to the configuration procedures needed to create an environment for e-science. The setup procedures to create, configure, and execute an application and then deallocate the environment are important aspects of cloud computing in science. The characteristics that should be evaluated are related to procedures and available tools to configure the environment. Features related to network configuration, time needed to create and configure VMs, and the hardware and software flexibility are also important. Criteria related to configuration procedures defined in our study are the following:

- **Setup procedures**: They consist of the user procedures to create and configure the environment in the cloud provider.
- **Hardware and software configurations**: These configurations are the available VMs size (number of cores and memory) and the capability to run different operating systems.
- **Network**: This criterion is related to the features offered by the provider to user access, as well as the interconnection between the VMs in the cloud.

- **Application porting procedures**: This consists of the adaptation that needs to be performed in the application for it to be executed in the cloud. The evaluation covers changes in both the source code and the execution environment.

To evaluate the performance of the cloud, it is necessary to compare it with a traditional system, which is a system whose performance the user knows and will be used as the basis for comparison. For a fair comparison, both the base and cloud systems need to present similar characteristics, mainly the number of cores of each system. The purpose is to have a direct comparison between a known system, the base system, and a new system, the cloud.

1.6 Analysis of Cloud Providers

1.6.1 Amazon Web Services

Amazon web services are one of the most widely known cloud providers. Many different kinds of services are offered, including storage, platform, and hosting services. Two of the most-used services of Amazon are the Amazon Elastic Compute Cloud (EC2) and Amazon Simple Storage Service (S3).

Amazon EC2 is an IaaS model and may be considered the central part of Amazon's cloud platform. It was designed to make web scaling easier for users. The interaction with the user is done through a web interface that permits obtaining and configuring any desired computing capacity with little difficulty. Amazon EC2 does not use regular configurations for the central processing unit (CPU) of instances available. Instead, it uses an abstraction called elastic compute units (ECUs). According to Amazon, each ECU provides the equivalent CPU capacity of a 1.0- to 1.2-GHz 2007 Opteron or 2007 Xeon processor. Amazon S3 is also an IaaS model and consists of a storage solution for the Internet. It provides storage through web service interfaces, such as REST and SOAP. There is no particular defined format of the stored objects; they are simple files. Inside the provider, the stored objects are organized into buckets, which are an Amazon proprietary method. The names of these buckets are chosen by the user, and they are accessible using a hypertext transfer protocol (HTTP) uniform resource locator (URL), with a regular web browser. This means that Amazon S3 can be easily used to replace static web hosting infrastructure. One example of an Amazon S3 user is the Dropbox service, provided as SaaS for the final user, with the user having a certain amount of storage in the cloud to store any desired file.

1.6.2 Rackspace

Rackspace was founded in 1998 as a typical hosting company with several levels of user support. The company developed the cloud services offered during company growth, and in 2009 they launched the Cloud Servers, which is a service of VMs and cloud files, an Internet-based service of storage.

The provider has data centers distributed in several regions: the United States, Europe, Australia, and Hong Kong. It is one of the major contributors of the Open Stack cloud project.

The product offered is the Open Cloud, which is an IaaS model. Several computing instances are provided that the user can launch and manage using a web-based control panel.

1.6.3 Microsoft Windows Azure

Microsoft started its initiative in cloud computing with the release of Windows Azure in 2008, which initially was a PaaS to develop and run applications written in the programming languages supported by the .NET framework. Currently, the company owns products that cover all types of service models. Online Services is a set of products that are provided as SaaS, while Windows Azure provides both PaaS and IaaS.

Windows Azure PaaS is a platform developed to provide the user the capability to develop and deploy a complete application into Microsoft's infrastructure. To have access to this service, the user needs to develop an application following the provided framework.

The Azure framework has support for a wide range of programming languages, including all .NET languages, Python, Java, and PHP. A generic framework is provided, in which the user can develop in any programming language that is supported by the Windows operating system (OS).

Windows Azure IaaS is a service developed to provide the user access to VMs running on Microsoft's infrastructure. The user has a set of base images of Windows and Linux OS, but other images can be created using Hyper-V. The user can also configure an image directly into Azure and capture it to use locally or to deploy to another provider that supports Hyper-V.

1.6.4 Google App Engine

Google App Engine (GAE) is a service that enables users to build and deploy their web applications on Google's infrastructure. The service model is PaaS, and the users of it are commonly developers. The users need to develop their application using the framework provided.

Currently, the languages supported are Python, Java, and Go. However, the provider intends to include more languages in the future.

The user develops and deploys the application using some of the available tool kits, and all the execution is managed by Google's staff. The high-availability and location distribution are automatically defined. Google is responsible for the elasticity, which is transparent to the user; this means that if one application receives many requests, the provider increases the resources, and the opposite also happens.

1.7 Cost Efficiency Evaluation

When the user decides to use a public cloud provider, it is necessary to calculate the cost efficiency [5] of this service and if it is better to use it or buy a cluster. To determine this, two calculations can be used, the cost efficiency factor and the break-even point [6].

1.7.1 Cost Efficiency Factor

To calculate the cost efficiency factor for different systems, two values are required. The first one is the cost of the cloud systems. This cost, in the great majority of cloud providers, is expressed as cost per hour. The second value is the overhead factor. To determine this factor, it is necessary to execute the same workload in all the candidate systems and in the base system.

The overhead factor O_F is the execution time in the candidate system ET_{CS} divided by the execution time in the base system ET_{BS}. The following equation represents this calculation:

$$O_F = \frac{ET_{CS}}{ET_{Bs}}$$

As an example, we want to compare a traditional server against a machine in the cloud. We define that the traditional server is the base system. We need to execute the same problem on both systems and then calculate the overhead factor. Assuming that the server takes 30 minutes to calculate and the cloud takes 60 minutes, applying the overhead factor equation, the result is 2 for the cloud. As the traditional system is the base system, its overhead factor is 1.

Using the overhead factor, it is possible to determine the cost efficiency factor CE_F. The cost efficiency factor is defined as the product between the cost per hour C_{HC} and the calculated overhead factor, resulting in the following equation:

$$CE_F = C_{HC} \times O_F$$

For example, using the calculated overhead factor 2 and assuming a cost per hour of $5.00 of a cloud machine, the resulting cost efficiency is $10.00 per hour. The cost efficiency gives the price to perform the same amount of work in the target system that the base system performs in 1 hour because the cost used in our equation is the cost per hour. If the result is less than the cost per hour of the base system, the candidate system presents a higher cost-benefit ratio than the base system. The cost efficiency factor also can be used to verify the scalability of the candidate system. If the number of machines increases and the cost efficiency factor is constant, the candidate system has the same scalability rate as the base system.

1.7.2 Break-Even Point

The break-even point, represented in Figure 1.2, represents the point at which the cost to use both the base and the candidate systems is the same, on a yearly basis. In a cloud computing environment, with its pay-per-use model, this metric is important. It represents the number of days in a year when it is cheaper to use a cloud instead of buying a server. Figure 1.2 represents the break-even point and is represented by the vertical bold line. If the user needs to use the system for fewer days than the break-even point (left side of the line), it is better to use a cloud, but if the usage is higher, it is more cost efficient to buy a server.

To calculate the break-even point, it is necessary to obtain the yearly cost of the base system. The yearly cost BS_{YC} represents the cost to maintain the system during a year; it is composed of the acquisition cost $Acq_\$$ of the machines themselves plus the maintenance costs $Ymn_\$$. To obtain the cost of the machines on a yearly basis, it is necessary to determine the usable lifetime LT of the machine, normally 3 to 5 years. It is necessary to divide the acquisition costs of the machines by the usage time; this calculation results in the cost per year of the machines. In the yearly cost, it is also necessary to include the maintenance, personnel, and facilities costs of the machines. The following equation calculates the yearly cost:

$$BS_{YC} = \frac{Acq_\$}{LT} + Ymn_\$$$

Using this value and the cost efficiency factor, we can determine the break-even point. The cost efficiency factor represents the cost on an hourly

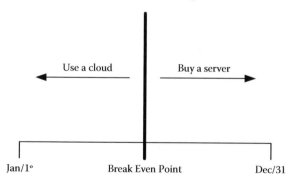

FIGURE 1.2
Break-even point.

basis; to obtain the number of days, the yearly cost is divided by the cost efficiency factor times 24. The following equation represents the break-even point calculation:

$$BEP = \frac{BS_{YC}}{CE_F \times 24}$$

where *BEP* represents the break-even point, BS_{YC} represents the calculated yearly cost of the base system, CE_F represents the cost efficiency factor, and 24 is the number of hours in a day. The result of this equation is expressed in number of days after which it becomes more cost efficient to use a server cluster instead of a cloud. It is important to remember that the number of days expressed by this equation is for continuous usage, 24 hours per day, but real-world usage is normally less than that. In a practical approach, if the server is used for fewer days per year than the break-even point, it is cheaper to use the cloud instead.

1.8 Evaluation of Providers: A Practical Example

To provide a better understanding of the proposed methodology, we will evaluate a hypothetical scenario. For this scenario, we need to execute the weather forecast for a region on a daily basis; the application is already developed in the Unix environment. Consider that we actually use a cluster to execute the application; now, this cluster needs to be changed because the supplier does not provide maintenance for it. We want to compare the acquisition of a new cluster to a public cloud provider to verify which presents the best solution in our case.

The first step is to verify if the application can be executed on both systems; because of the Unix execution model, it is compatible with the new cluster and with the cloud since both have a compatible operating system. The cloud provides adequate tools to create a cluster-like environment to execute parallel applications, and the delivery procedures are performed using standard network protocols, such as FTP (file transfer protocol). The conclusion is that the application can be executed both on the new cluster and in the cloud.

The second step is related to the performance of the solutions; it is necessary to execute the same workload on both and then calculate the overhead, in terms of execution time, of the solutions. The workload in our example is the weather forecast application itself, with real input data, and we assume the cluster as the base system and the cloud as the candidate system. The execution time for the cluster was 4 hours (240 minutes), and the execution time for the cloud was 6 hours (360 minutes). Applying the overhead factor equation, we have the following result:

$$\frac{360}{240} = 1.5$$

which means that the overhead factor to execute the same calculation in the cloud, compared to the cluster, is 1.5. In other words, the time to execute the same application with the same data in the cloud takes 50% more time than the cluster. The weather forecast needs to be executed daily in less than 12 hours; therefore, both solutions present adequate execution time.

The third and final step is related to the economic evaluation of both solutions. The first input for this calculation is the price of both solutions. The acquisition cost of the cluster is $1.3 million, and it will be used during its lifetime of 10 years. To maintain the cluster, it is necessary to contract a maintenance specialist for $3,000 per month, or $36,000 per year. Moreover, the energy consumption of this system is $1,000 per month or $12,000 per year. With all these costs, we can use the yearly cost equation; the results are

$$\frac{\$1,300,000}{10} + \$48,000 = \$178,000$$

This result means that the cost per year with the cluster is $178,000; this value will be used in the break-even point assessment. Another component of the break-even point is the cost efficiency factor, assuming a cost per hour of $50.00 for the cloud machine. Using the calculated overhead factor of 1.5, the resulting cost efficiency factor for the cloud is 75.00 ($/hour). Using both the yearly cost and the cost efficiency factor, we can determine the break-even point with the following calculation:

$$\frac{178,000}{75 \times 24} = 98.88 \,(days)$$

but this result is related to full usage for 24 hours a day. The real usage of the cloud will be 6 hours a day, which is the time required to perform the weather forecast for our city. Then, we can adjust the break-even point calculation for 6 hours; the new result is

$$\frac{178,000}{75 \times 6} = 395.55 \,(days)$$

This result is interpreted to mean that the number of days when the cloud has a better cost-benefit ratio than the cluster is 395 days in a year. We can conclude that the use of a cloud instead of a cluster is cheaper.

1.9 Conclusions

In the discussion in this chapter, with the focus on economic viability, we can conclude that the cloud computing model is a competitive alternative to be used for e-science applications. The recommended configuration is the public implementation model, by which the user pays according to the use of the application.

Moreover, due to the cost efficiency evaluation model presented, it is possible to determine when using a cloud is better in terms of cost-benefit ratio than to buy a physical server. This metric can be used during the decision process regarding which platform will be used to create the e-science service.

References

1. C. Ward, N. Aravamudan, K. Bhattacharya, K. Cheng, R. Filepp, R. Kearney, B. Peterson, L. Shwartz, and C. Young. Workload migration into clouds—challenges, experiences, opportunities. In *Cloud Computing (CLOUD), 2010 IEEE 3rd International Conference on*, July 2010, pp. 164–171.

2. W. Lu, J. Jackson, and R. Barga. Azureblast: a case study of developing science applications on the cloud. In *Proceedings of the 19th ACM International Symposium on High Performance Distributed Computing, ser. HPDC '10*. New York: ACM, 2010, pp. 413–420.

3. P. Mell and T. Grance. *The NIST Definition of Cloud Computing*. Tech. Rep. 2011. http://www.mendeley.com/research/the-nist-definition-about-cloud-computing/.

4. E. Roloff, F. Birck, M. Diener, A. Carissimi, and P. O. A. Navaux. Evaluating high performance computing on the Windows Azure platform. In *Proceedings of the 2012 IEEE 5th International Conference on Cloud Computing (CLOUD 2012)*, 2012, pp. 803–810.

5. D. Kondo, B. Javadi, P. Malecot, F. Cappello, and D. Anderson. Cost-benefit analysis of cloud computing versus desktop grids. In *Parallel Distributed Processing, 2009. IPDPS 2009. IEEE International Symposium on*, May 2009, pp. 1–12.

6. E. Roloff, M. Diener, A. Carissimi, and P. O. A. Navaux. High performance computing in the cloud: deployment, performance and cost efficiency. In *Proceedings of the 2012 IEEE 4th International Conference on Cloud Computing Technology and Science (CLOUDCOM)*, 2012, pp. 371–378.

2

Cloud-Based Infrastructure for Data-Intensive e-Science Applications: Requirements and Architecture

Yuri Demchenko, Canh Ngo, Paola Grosso, Cees de Laat, and Peter Membrey

CONTENTS

Summary

This chapter discusses the challenges that are imposed by big data on the modern and future e-scientific data infrastructure (SDI). The chapter discusses the nature and definition of big data, including such characteristics as volume, velocity, variety, value, and veracity. The chapter refers to different scientific communities to define requirements on data management, access control, and security. The chapter introduces the scientific data life cycle management (SDLM) model, which includes all the major stages and reflects specifics in data management in modern e-science. The chapter proposes the generic SDI architectural model that provides a basis for building interoperable data or project-centric SDI using modern technologies and best practices. The chapter discusses how the proposed models SDLM and SDI can be naturally implemented using modern cloud-based infrastructure services and analyses security and trust issues in cloud-based infrastructure and summarizes requirements to access control and access control infrastructure that should allow secure and trusted operation and use of the SDI.

2.1 Introduction

The emergence of data-intensive science is a result of modern science computerization and an increasing range of observations, experimental data collected from specialist scientific instruments, sensors, and simulation in every field of science. Modern science requires wide and cross-border research collaboration. The e-science scientific data infrastructure (SDI) needs to provide an environment capable of both dealing with the ever-increasing heterogeneous data production and providing a trusted collaborative environment for distributed groups of researchers and scientists. In addition, SDI needs on the one hand to provide access to existing scientific information, including that in libraries, journals, data sets, and specialist scientific databases and on the other hand to provide linking between experimental data and publications.

Industry is also experiencing wide and deep technology refactoring to become data intensive and data powered. Cross-fertilization between emerging data-intensive/-driven e-science and industry will bring new data-intensive technologies that will drive new data-intensive/-powered applications.

Further successful technology development will require the definition of the SDI and overall architecture framework of data-intensive science. This will provide a common vocabulary and allow concise technology evaluation and planning for specific applications and collaborative projects or groups.

Big data technologies are becoming a current focus and a new "buzzword" both in science and in industry. Emergence of big data or data-centric technologies indicates the beginning of a new form of continuous technology advancement that is characterized by overlapping technology waves related to different aspects of human activity from production and consumption to collaboration and general social activity. In this context, data-intensive science plays a key role.

Big data are becoming related to almost all aspects of human activity, from just recording events to research, design, production, and digital services or products delivery, to the final consumer. Current technologies, such as cloud computing and ubiquitous network connectivity, provide a platform for automation of all processes in data collection, storing, processing, and visualization.

Modern e-science infrastructures allow targeting new large-scale problems whose solution was not possible previously (e.g., genome, climate, global warming). e-Science typically produces a huge amount of data that need to be supported by a new type of e-infrastructure capable of storing, distributing, processing, preserving, and curating these data [1, 2]: We refer to these new infrastructures as the SDI.

In e-science, the scientific data are complex multifaceted objects with complex internal relations. They are becoming an infrastructure of their own and need to be supported by corresponding physical or logical infrastructures to store, access, process, visualize, and manage these data.

The emerging SDI should allow different groups of researchers to work on the same data sets, build their own (virtual) research and collaborative environments, safely store intermediate results, and later share the discovered results. New data provenance, security, and access control mechanisms and tools should allow researchers to link their scientific results with the initial data (sets) and intermediate data to allow future reuse/repurposing of data (e.g., with the improved research technique and tools).

This chapter analyzes new challenges imposed on modern e-science infrastructures by the emerging big data technologies; it proposes a general approach and architecture solutions that constitute a new scientific data life cycle management (SDLM) model and the generic SDI architecture model that provides a basis for heterogeneous SDI component interoperability and integration, in particular based on cloud infrastructure technologies.

The chapter is primarily focused on SDI; however, it provides analysis of the nature of big data in e-science, industry, and other domains; analyses their commonalities and differences; and discusses possible cross-fertilization between two domains.

The chapter refers to ongoing research on defining the big data infrastructure for e-science initially presented elsewhere [3, 4] and significantly extends it with new results and a wider scope to investigate relations between big data technologies in e-science and industry. With a long tradition of working with a constantly increasing volume of data, modern science can offer industry scientific analysis methods, while industry can bring big data technologies and tools to wider public access.

The chapter is organized as follows: Section 2.2 looks into the definition and nature of big data in e-science, industry, business, and social networks, also analyzing the main drivers for big data technology development. Section 2.3 gives an overview of the main research communities and summarizes requirements for future SDI. Section 2.4 discusses challenges to data management in big data science, including a discussion of SDLM. Section 2.5 introduces the proposed e-SDI architecture model that is intended to answer the future big data challenges and requirements. Section 2.6 discusses SDI implementation using cloud technologies. Section 2.6 discusses security and trust-related issues in handling data and summarizes specific requirements to access the control infrastructure for modern and future SDIs.

2.2 Big Data Definition

2.2.1 Big Data in e-Science, Industry, and Other Domains

Science traditionally has dealt with challenges to handle large volumes of data in complex scientific research experiments. Scientific research typically includes a collection of data in passive observation or active experiments that aim to verify one or another scientific hypotheses. Scientific research and discovery methods typically are based on the initial hypothesis and a model that can be refined based on the collected data. The refined model may lead to a new, more advanced and precise experiment or reevaluation of the previous data. Another distinctive feature of modern scientific research is that it suggests wide cooperation between researchers to challenge complex problems and run complex scientific instruments.

In industry, private companies will not share data or expertise. When dealing with data, companies will always intend to keep control over their information assets. They may use shared third-party facilities, like clouds, but special measures need to be taken to ensure data protection, including data sanitization. Also, companies might use shared facilities only for proof of concept and do production data processing at private facilities. In this respect, we need to accept that science and industry cannot be done in the same way; consequently, this will be reflected in how they can interact and how the big data infrastructure and tools can be built.

With the proliferation of digital technologies into all aspects of business activities and emerging big data technologies, the industry is entering a new playing field when it needs to use scientific methods to benefit from the possibility of collecting and mining data for desirable information, such as market prediction, customer behavior predictions, social groups activity predictions, and so on.

A number of discussions and blog articles [5–7] suggested that the big data technologies need to adopt scientific discovery methods that include iterative model improvement and collection of improved data and reuse of collected data with an improved model.

According to a blog article by Mike Gualtieri from Forrester [7]: "Firms increasingly realize that [big data] must use predictive and descriptive analytics to find nonobvious information to discover value in the data. Advanced analytics uses advanced statistical, data mining and machine learning algorithms to dig deeper to find patterns that you can't see using traditional BI tools, simple queries, or rules."

2.2.2 The Big Data Definition

Despite the fact that the term *big data* has become a new buzzword, there is no consistent definition for big data or detailed analysis of this new emerging technology. Most discussions until now have been in the blogosphere, where the most significant big data characteristics have been identified and been commonly accepted [8–10]. In this section, we summarize available definitions and propose a consolidated view of the generic big data features that would help us define requirements to support big data infrastructure, particularly the SDI.

As a starting point, we can refer to a simple definition [9]: "Big Data: a massive volume of both structured and unstructured data that is so large that it's difficult to process using traditional database and software techniques." A related definition of the data-intensive science is given in the book *The Fourth Paradigm: Data-Intensive Scientific Discovery* by the computer scientist Jim Gray [10]: "The techniques and technologies for such data-intensive science are so different that it is worth distinguishing data-intensive science from computational science as a new, fourth paradigm for scientific exploration" (p. xix).

2.2.3 Five Vs of Big Data

In a number of discussions and articles, big data are attributed to have such native generic characteristics as volume, velocity, and variety, also referred to as the "3 Vs of big data." After being stored and entered into the processing stages or workflow, big data acquire new properties, value and veracity, which together constitute the five Vs of big data: volume, velocity, variety, value, and veracity [4]. Figure 2.1 illustrates the features related to the 5 Vs, which are analyzed next.

2.2.3.1 Volume

Volume is the most important and distinctive feature of big data that imposes additional and specific requirements for all traditional technologies

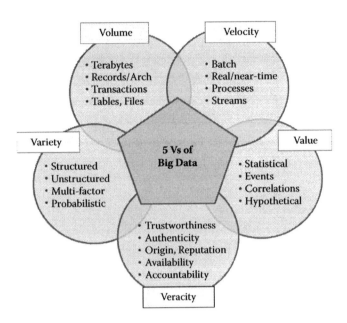

FIGURE 2.1
Five Vs of big data.

and tools currently used. In e-science, growth of the data amount is caused by advancements in both scientific instruments and SDI. In many areas, the trend is actually to include data collections from all observed events, activities, and sensors, which became possible and is important for social activities and social sciences.

Big Data volume includes such features as size, scale, amount, and dimension for tera- and exascale data recording either data-rich processes or data collected from many transactions and stored in individual files or databases. All need to be accessible, searchable, processed, and manageable.

Two examples from e-science also provide different characteristics of data and different processing requirements:

- The Large Hadron Collider (LHC) [11, 12] produces on average 5 PB (petabytes) of data a month that are generated in a number of short collisions that make them unique events. The collected data are filtered, stored, and extensively searched for single events that may confirm a scientific hypothesis.
- The LOFAR (Low-Frequency Array) [13] is a radio telescope that collects about 5 PB every hour; however, the data are processed by a correlator, and only correlated data are stored.

In industry, global services providers such as Google [14], Facebook [15], and Twitter [16] are producing, analyzing, and storing data in huge amounts as regular activity/production services. Although some of their tools and processes are proprietary, they actually prove the feasibility of solving big data problems at the global scale and significantly push the development of the Open Source big data tools.

2.2.3.2 Velocity

Big data are often generated at high speed, including data generated by arrays of sensors or multiple events; these data need to be processed in real time or near real time, in a batch, or as streams (e.g., for visualization). As an example, the LHC ATLAS detector [12] uses about 80 readout channels and collects up to 1 PB of unfiltered data per second, which are reduced to approximately 100 MB per second. This should record up to 40 million collision events per second.

Industry can also provide numerous examples when data registration, processing, or visualization imposes similar challenges.

2.2.3.3 Variety

Variety deals with the complexity of big data and information and semantic models behind these data. This results in data collected as structured, unstructured, semistructured, and mixed data. Data variety imposes new requirements for data storage and database design, which should have dynamic adaptation to the data format, particularly scaling up and down.

Biodiversity research [17] provides a good example of the data variety that is a result of the collection and processing of information from a wide range of sources and the relation of the collected information to species population, genomic data, climate, satellite information, and more. Another example can be urban environment monitoring (also called "smart cities" [18]), which requires operating, monitoring, and evolving numerous processes, individuals, and associations.

Adopting data technologies in traditionally non-computer-oriented areas such as psychology and behavior research, history, and archeology will generate especially rich data sets.

2.2.3.4 Value

Value is an important feature of the data that is defined by the added value that the collected data can bring to the intended process, activity, or predictive analysis/hypothesis. Data value will depend on the events or processes the data represent, such as processes that are stochastic, probabilistic, regular, or random. Depending on this, requirements may be imposed to collect all data, store the data for a longer period (for some possible event of interest), and so on. In this respect, data value is closely related to the data volume and variety. The stock exchange financial data provide a good example of high-volume data that have high value for real-time market trend monitoring, but decreasing value with time and market volatility [19].

2.2.3.5 Veracity

The veracity dimension of big data includes two aspects: data consistency (or certainty), which can be defined by statistical reliability of the data, and data trustworthiness, which is defined by a number of factors, among them data origin and collection and processing methods, including trusted infrastructure and facility.

Big data veracity ensures that the data used are trusted, authentic, and protected from unauthorized access and modification. The data must be secured during their whole life cycle, from collection from trusted sources to processing on trusted computing facilities and storage on protected and trusted storage facilities.

The following aspects define and need to be addressed to ensure data veracity:

- Integrity of data and linked data (e.g., for complex hierarchical data, distributed data)
- Data authenticity and (trusted) origin
- Identification of both data and source
- Computer and storage platform trustworthiness
- Availability and timeliness
- Accountability and reputation

Data veracity relies entirely on the security infrastructure deployed and available from the big data infrastructure [20].

2.3 Research Infrastructures and Infrastructure Requirements

This section refers to and provides a short overview of different scientific communities, in particular as defined by the European Research Area (ERA) [21], to define requirements for the infrastructure facility, data-processing and management functionalities, user management, access control, and security.

2.3.1 Paradigm Change in Modern e-Science

Modern e-science is moving to the data-intensive technologies that are becoming a new technology driver and require rethinking a number of infrastructure architecture and operational models, components, solutions, and processes to address the following general challenges [2, 4]:

- Exponential growth of data volume produced by different research instruments or collected from sensors
- Need to consolidate e-infrastructures as persistent research platforms to ensure research continuity and cross-disciplinary collaboration, deliver/offer persistent services, with an adequate governance model

The recent advancements in the general computer and big data technologies facilitate the paradigm change in modern e-science that is characterized by the following features:

- Automation of all e-science processes, including data collection, storing, classification, indexing, and other components of the general data curation and provenance
- Transformation of all processes, events, and products into digital form by means of multidimensional, multifaceted measurements, monitoring, and control; digitizing existing artifacts and other content
- Possibility of reusing the initial and published research data with possible data repurposing for secondary research
- Global data availability and access over the network for a cooperative group of researchers, including wide public access to scientific data
- Existence of necessary infrastructure components and management tools that allow fast infrastructures and services composition, adaptation and provisioning on demand for specific research projects and tasks
- Advanced security and access control technologies that ensure secure operation of the complex research infrastructures and scientific instruments and allow creating a trusted secure environment for cooperating groups and individual researchers

The future SDI should support the whole data life cycle and explore the benefit of data storage/preservation, aggregation, and provenance on a large scale and during long or unlimited periods of time. It is important that this infrastructure ensure data security (integrity, confidentiality, availability, and accountability) and data ownership protection. With current needs to process big data that require powerful computation, there should be a possibility of enforcing data/data set policy so that they can be processed on trusted systems or comply with other requirements. Researchers must trust the SDI to process their data on SDI facilities and be assured that their stored research data are protected from nonauthorized access. Privacy issues also arise from the distributed remote character of SDI, which can span multiple countries with different local policies. This should be provided by the corresponding access control and accounting infrastructure (ACAI), which is an important component of SDI [20, 22].

2.3.2 Research Communities and Specific SDI Requirements

A short overview of some research infrastructures and communities, particularly the ones defined for the ERA [21], allows a better understanding of specific requirements for the future SDIs that are capable of addressing big data challenges. Existing studies of European e-infrastructures analyzed the scientific communities' practices and requirements; examples of these studies are those undertaken by the SIENA Project [23], EIROforum Federated Identity Management Workshop [24], European Grid Infrastructure (EGI) Strategy Report [25], and UK Future Internet Strategy Group Report [26].

The high-energy physics (HEP) community represents a large number of researchers, unique expensive instruments, and a huge amount of data that are generated and need to be processed continuously. This community already has the operational Worldwide LHC Computing Grid (WLCG) [11] infrastructure to manage and access data, protect their integrity, and support the whole scientific data life cycle. WLCG development was an important step in the evolution of European e-infrastructures that currently serve multiple scientific communities in Europe and internationally. The EGI cooperation [27] manages European and worldwide infrastructure for HEP and other communities.

Material science and analytical and low-energy physics (proton, neutron, laser facilities) are characterized by short projects and experiments and consequently a highly dynamic user community. A highly dynamic supporting infrastructure and advanced data management infrastructure to allow wide data access and distributed processing are needed.

The environmental and earth science community and projects target regional or national and global problems. Huge amounts of data are collected from land, sea, air, and space and require an ever-increasing amount of storage and computing power. This SDI requires reliable fine-grained access control to huge data sets, enforcement of regional issues, and policy-based data filtering (data may contain national security-related information) while tracking data use and maintaining data integrity.

Biological and medical sciences (also defined as life sciences) have a general focus on health, drug development, new species identification, and new instrument development. They generate a massive amount of data and new demands for computing power, storage capacity, and network performance for distributed processes, data sharing, and collaboration. Biomedical data (health care, clinical case data) are privacy-sensitive data and must be handled according to the European policy on processing of personal data [27].

Biodiversity research [17] involves research data and research specialists from at least biology and environmental research and may include data about climate, weather, and satellite observation. This primarily presents challenges for not only integrating different sources of information with different data models and processing a huge amount of collected information but also may require fast data processing in case of natural disasters. The projects LifeWatch [28] and ENVRI (Common Operations of Environmental Research

Infrastructure) [29] present good examples of which research approaches and what kind of data are used.

Social science and humanities communities and projects are characterized by multilateral and often global collaborations between researchers who need to be engaged into collaborative groups or communities and supported by collaborative infrastructure to share data and discovery/research results and cooperatively evaluate results. The current trend to digitize all currently collected physical artifacts will create in the near future a huge amount of data that must be widely and openly accessible.

2.3.3 General SDI Requirements

From the overview, we can extract the following general infrastructure requirements for SDI for emerging big data science:

- Support for long-running experiments and a large volume of heterogeneous data generated at high speed
- On-demand infrastructure provisioning to support data sets and scientific workflows and mobility of data-centric scientific applications
- Provision of high-performance computing facilities to allow complex data analytics with evolving research models
- Support for distributed and mobile sensor networks for observation data collection and advance information visualization
- Support for virtual scientists' communities, addressing dynamic user groups creation and management, federated identity management
- Support for the whole data life cycle management, particularly advanced data provenance, data archiving, and consistent data identification
- Multitier interlinked data distribution and replication
- Provision of a trusted environment for data storage and processing
- Support for data integrity, confidentiality, accountability
- Policy binding to data to protect privacy, confidentiality, and intellectual property rights (IPR)

2.4 Scientific Data Management

2.4.1 Scientific Information and Data in Modern e-Science

Emergence of computer-aided research methods is transforming the way research is done and scientific data are used. The following types of scientific data are defined and illustrated in a form of scientific data pyramid (see Figure 2.2) [22]:

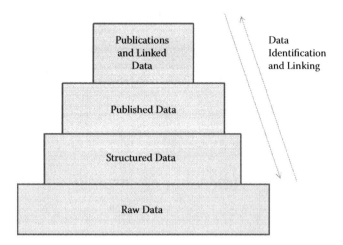

FIGURE 2.2
Scientific data pyramid.

- Raw data collected from observations and from experiments (what actually is done according to an initial research model or hypothesis).

- Structured data and data sets that went through data filtering and processing (supporting some particular formal model, which is typically refined from the initial model). These data are already stored in repositories and may be shared with collaborative groups of researchers.

- Published data that support one or another scientific hypothesis, research result, or statement. These data are typically linked to scientific publications as supplemental materials; they may be located on the publisher's platform or authors' institution platform and have open or licensed access.

- Data linked and embedded into publications to support wide research consolidation, integration, and openness.

Once the data are published, it is essential to allow other scientists to be able to validate and reproduce the data in which they are interested and possibly contribute new results. Capturing information about the processes involved in transformation from raw data until the generation of published data becomes an important aspect of scientific data management. Scientific data provenance becomes an issue that also needs to be taken into consideration by SDI providers [30].

Another aspect to take into consideration is to guarantee reusability of published data within the scientific community. Understanding the semantics of the published data becomes an important issue to allow for reusability; this traditionally has been done manually. However, as we anticipate an unprecedented scale of published data that will be generated in big data science, attaching a clear data semantic becomes a necessary condition for efficient reuse of published data. Learning from best practices in the semantic web community on how to provide reusable published data will be one consideration that will be addressed by SDI.

Big data are typically distributed both on the collection side and on the processing/access side: Data need to be collected (sometimes in a time-sensitive way or with other environmental attributes), distributed, or replicated. Linking distributed data is one of the problems to be addressed by SDI.

The European Commission's initiative to support open access to scientific data from publicly funded projects suggests introduction of the following mechanisms to allow linking publications and data [31]:

- PID: persistent data ID [32]
- ORCID: Open Researcher and Contributor Identifier [33].

2.4.2 Data Life Cycle Management in Scientific Research

e-Science enabled by computers and information technology (IT) allows multipurpose data collection and use and advanced data processing. A possibility to store the initial data sets and all intermediate results will allow for future data use, in particular data repurposing and secondary research, as the technology and scientific methods develop.

Emergence of computer-aided research methods is transforming the way research is performed and scientific data are processed or used. This is also reflected in the changed SDLM shown in Figure 2.3 and discussed next.

We refer to the extensive study of the SDLM models [34]. The traditional scientific data life cycle includes a number of stages (see Figure 2.3a):

- Research project or experiment planning
- Data collection
- Data integration and processing
- Research result publication
- Discussion, feedback
- Archiving (or discarding)

The new SDLM model requires data storage and preservation at all stages, which should allow data reuse or repurposing and secondary research on the processed data and published results. However, this is possible only if the full data identification, cross-reference, and linkage are implemented in the SDI. Data integrity, access control, and accountability must be supported during the entire data life cycle. Data curation is an important component of the discussed SDLM and must also be done in a secure and trustworthy way.

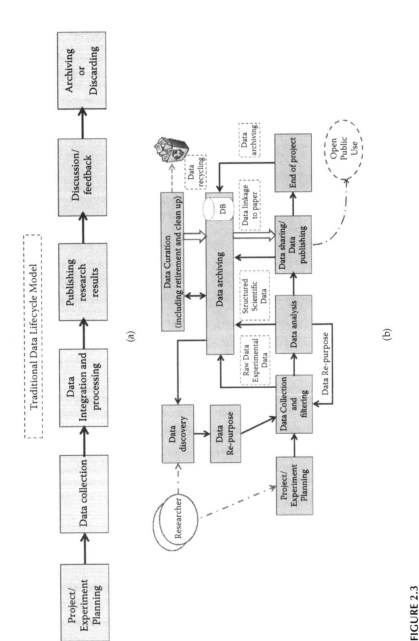

FIGURE 2.3
(a) Scientific data life cycle management in traditional science. (b) Scientific data life cycle management in e-science.

The following support data security and access control to scientific data during their life cycle: data acquisition (experimental data), initial data filtering, specialist processing, research data storage and secondary data mining, data and research information archiving.

2.5 Scientific Data Infrastructure Architecture Model

The proposed generic SDI architecture model provides a basis for building interoperable data or project-centric SDI using modern technologies and best practices. Figure 2.4 shows the multilayer SDI architecture for e-science (e-SDI) that contains the following layers:

Layer D1: Network infrastructure layer represented by either the general-purpose Internet infrastructure or dedicated network infrastructure

Layer D2: Data centers and computing resources/facilities

Layer D3: Infrastructure virtualization layer represented by the cloud/grid infrastructure services and middleware supporting specialized scientific platform deployment and operation

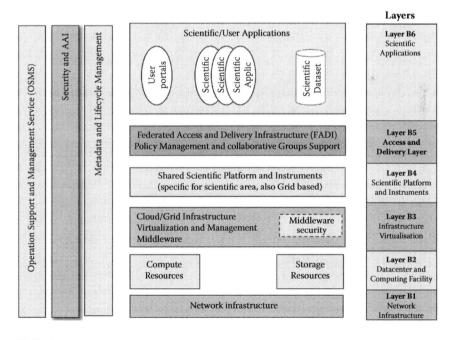

FIGURE 2.4
The proposed SDI architecture model.

Layer D4: (Shared) scientific platforms and instruments specific for different research areas.

Layer D5: Access and delivery layer that represents the general Federated Access and Delivery Infrastructure (FADI) that includes infrastructure components for interconnecting, integrating, and operating complex scientific infrastructure to support project-oriented collaborative groups of researchers

Layer D6: Scientific applications, subject-specific databases, and user portals/clients

Note: The D prefix denotes the relation to the data infrastructure.

We also define the three cross-layer planes: operational support and management system, security plane, and metadata and life cycle management.

The dynamic character of SDI and its support of distributed multifaceted communities are guaranteed by the following dedicated layers: D3, the infrastructure virtualization layer that typically uses modern cloud technologies, and D5, the FADI layer that incorporates related federated infrastructure management and access technologies [21, 35, 36]. Introduction of the FADI layer reflects current practices in building and managing complex SDIs (and enterprise infrastructures) and allows independently managed infrastructures to share resources and support interorganizational cooperation.

Network infrastructure is presented as a separate lower layer in e-SDI, but dedicated network infrastructure provisioning is also relevant to the FADI layer. Network aspects in big data are becoming even more important than for computer grids and clouds. We can identify two main challenges that big data transport will impose on the underlying layer of the SDI:

- Timely delivery to bring all data where required with the smallest possible latency
- Cost reduction to optimize the amount of network equipment required (either via purchasing it or on a pay-per-use basis) without scarifying the quality of service (QoS)

For many SDIs, the basic best-effort Internet is the only available network transport architecture. In these cases, given the constraints imposed by this shared medium, it will be difficult to fully provide the low latency and guaranteed delivery required for big data processing. Performance may be lower, but it will be manageable. Fewer SDIs will rely on circuit-based networks, for which the timely delivery of data will be guaranteed but the costs for operating or using the network path will be significantly higher.

We see a third possibility for dealing with big data at the lowest layer of the SDI. Emerging protocols for network programmability (e.g., OpenFlow and in general software-defined networks) provide interesting solutions. By fully controlling the network equipment, both time and costs can be optimized.

Although the dilemma of moving data to computing facilities or moving computing to the data location can be solved in some particular cases, processing highly distributed data on MPP (massively parallel processing) infrastructures will require a special design of the internal MPP network infrastructure.

2.6 Cloud-Based Infrastructure Services for SDI

Figure 2.5 illustrates the typical e-science or enterprise collaborative infrastructure that is created on demand and includes enterprise proprietary and cloud-based computing and storage resources, instruments, control and monitoring system, visualization system, and users represented by user clients and typically residing in real or virtual campuses. The main goal of the enterprise or scientific infrastructure is to support the enterprise or scientific workflow and operational procedures related to process monitoring and data processing. Cloud technologies simplify the building of such infrastructure and provision it on demand. Figure 2.5 illustrates how an example enterprise or scientific workflow can be mapped to cloud-based services and later deployed and operated as an instant intercloud infrastructure. It contains cloud infrastructure segments IaaS (infrastructure as a service)

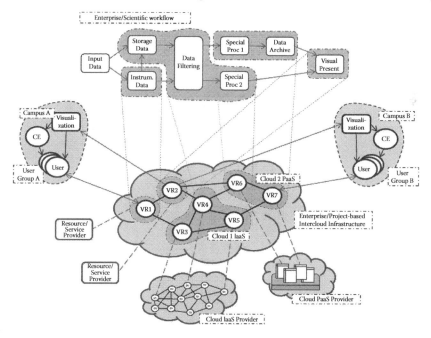

FIGURE 2.5
From scientific workflow to cloud-based infrastructure.

(VR3–VR5) and PaaS (platform as a service) (VR6, VR7); separate virtualized resources or services (VR1, VR2); two interacting campuses, A and B; and interconnecting them to a network infrastructure that in many cases may need to use dedicated network links for guaranteed performance.

Efficient operation of such infrastructure will require both overall infrastructure management and individual services and infrastructure segments to interact between themselves. This task is typically out of the scope of the existing cloud service provider models but will be required to support perceived benefits of the future cloud-based e-SDI. These topics are a subject for us in other research on the intercloud architecture framework (ICAF) [37–39]. The ICAF provides a common basis for building adaptive and on-demand provisioned multiprovider cloud-based infrastructure services.

Besides the general cloud-based infrastructure services (storage, compute, infrastructure/virtual machine [VM] management), the following specific applications and services are required to support big data and other data-centric applications [40]:

- Cluster services
- Hadoop-related services and tools
- Specialist data analytics tools (logs, events, data mining, etc.)
- Databases/servers SQL, NoSQL
- MPP databases
- Big data management tools
- Registries, indexing/search, semantics, namespaces
- Security infrastructure (access control, policy enforcement, confidentiality, trust, availability, privacy)
- Collaborative environment (groups management)

Big data analytics tools are currently offered by the major cloud services providers, such as Amazon Elastic MapReduce and Dynamo [41], Microsoft Azure HDInsight [42], IBM Big Data Analytics [43]. HPCC Systems by LexisNexis [44], Scalable Hadoop, and data analytics tools services are offered by a few companies that position themselves as big data companies, such as Cloudera [45] and a few others [46].

2.7 Security Infrastructure for Big Data

2.7.1 Security and Trust in Cloud-Based Infrastructure

Ensuring data veracity in big data infrastructure and applications requires deeper analysis of all factors affecting data security and trustworthiness

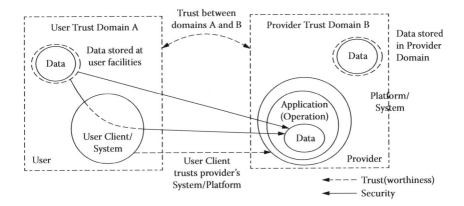

FIGURE 2.6
Security and trust in data services and infrastructure.

during their whole life cycle. Figure 2.6 illustrates the main actors and their relations when processing data on a remote system. User or customer and service provider are the two actors concerned with their own data and content security and each other's system/platform trustworthiness: The user wants to be sure that his or her data are secure when processed or stored on the remote system.

Figure 2.6 illustrates the complexity of trust and security relations even in a simple use case of the direct user/provider interaction. In clouds, data security and the trust model need to be extended to a distributed, multi-domain, and multiprovider environment. In the general case of a multi-provider and multitenant e-science cooperative environment, the e-SDI security infrastructure should support on-demand created and dynamically configured user groups and associations, potentially reusing existing experience in managing virtual organizations (VOs) and VO-based access control in computer grids [47, 48].

Data-centric security models, when used in a generically distributed and multiprovider e-SDI environment, will require policy binding to data and a fine-grained data access policy that should allow flexible policy definition based on the semantic data model. Based on our experience, the XACML (eXtensible Access Control Mark-up Language) policy language can provide a good basis for such functionality [49, 50]. However, support of the data life cycle and related provenance information will require additional research in policy definition and underlying trust management models.

2.7.2 General Requirements for a Federated Access Control Infrastructure

To support both secure data processing and project based collaboration of researchers, the future SDI should be supported by a corresponding

Federated Access Control Infrastructure (FACI) that would ensure normal infrastructure operation and assets and information protection and allow user authentication and policy enforcement in the distributed multi-organization environment. The future SDI should support the entire data life cycle and explore the benefits of data storage/preservation, aggregation, and provenance on a large scale and during a long or unlimited period of time; accordingly, the future FACI should support all stages of the data life cycle, including policy attachment to data to ensure persistency of the data policy enforcement during continuous online and offline processes.

The required FACI should support the following features of the future SDI:

- Empower researchers (and garner their trust) to do their data processing on shared facilities of large data centers with guaranteed data and information security.

- Motivate/assure researchers that they can share or open their research environment to other researchers by providing tools for instantiation of customized preconfigured infrastructures to allow other researchers to work with existing or their own data sets.

- Protect data policy, ownership, linkage (with other data sets and newly produced scientific/research data) when providing (long term) data archiving. Data preservation technologies should themselves ensure data readability and accessibility with the changing technologies.

2.8 Summary and Future Development

The information presented in this chapter provides a snapshot of the fast-developing big data and data analytics technologies that merge modern e-science research methods and experience of dealing with the large-scale problems, on one hand, and modern industry speed of technology development and global scale of implementation and services availability on the other. At this stage, we summarized and presented rethinking on some widely used definitions related to big data; further research will require a more formal approach and taxonomy of the general big data use cases in both science and industry.

As a part of general infrastructure research, we will continue research on the infrastructure issues in big data, targeting a more detailed and technology-oriented definition of SDI and related security infrastructure definition. Special attention will be given to defining the whole cycle of the provisioning of SDI services on demand, specifically tailored to support instant scientific workflows using cloud IaaS and PaaS platforms. This research will also be supported by development of the corresponding cloud architecture framework and ICAF to support the big data e-science processes and infrastructure operation. Particular focus will be made on the federated cloud and intercloud service provisioning model.

Although the currently proposed SDLM definition has been accepted as the European Commission Study recommendation [21], the further definition of the related metadata, procedures, and protocols as well as SDLM extension to the general big data life cycle is required.

The research presented is planned to contribute to the two standardization bodies related to the emerging big data technology with which authors are actively involved: the Research Data Alliance (RDA) [51] and the recently established NIST Big Data Working Group (NBD-WG) [52].

References

1. Thanos, C. Global Research Data Infrastructures: towards a 10-year vision for global research data infrastructures. Final Roadmap, March 2012. http://www.grdi2020.eu/Repository/FileScaricati/6bdc07fb-b21d-4b90-81d4-d909fdb96b87.pdf.

2. Riding the wave: how Europe can gain from the rising tide of scientific data. Final report of the High Level Expert Group on Scientific Data. October 2010. http://cordis.europa.eu/fp7/ict/e-infrastructure/docs/hlg-sdi-report.pdf.

3. Demchenko, Y., Z. Zhao, P. Grosso, A. Wibisono, and C. de Laat. Addressing big data challenges for scientific data infrastructure. Presented at the 4th IEEE Conference on Cloud Computing Technologies and Science (CloudCom2012), December 3–6, 2012, Taipei, Taiwan.

4. Demchenko, Y., P. Membrey, P. Grosso, and C. de Laat. Addressing big data issues in scientific data infrastructure. Presented at the First International Symposium on Big Data and Data Analytics in Collaboration (BDDAC 2013). Part of the 2013 International Conference on Collaboration Technologies and Systems (CTS 2013), May 20–24, 2013, San Diego, CA, USA.

5. *Reflections on Big Data, Data Science and Related Subjects.* Blog by Irving Wladawsky-Berger. http://blog.irvingwb.com/blog/2013/01/reflections-on-big-data-data-science-and-related-subjects.html.

6. Gantz, J., and D. Reinsel. Extracting value from chaos. *IDC IVIEW.* June 2011. http://www.emc.com/collateral/analyst-reports/idc-extracting-value-from-chaos-ar.pdf.

7. The Forrester Wave: Big Data Predictive Analytics Solutions, Q1 2013. Mike Gualtieri, January 13, 2013. http://www.forrester.com/pimages/rws/reprints/document/85601/oid/1-LTEQDI.

8. Dumbill, E. What is big data? An introduction to the big data landscape. http://strata.oreilly.com/2012/01/what-is-big-data.html.

9. The Big Data Long Tail. Blog post by Jason Bloomberg on January 17, 2013. http://www.devx.com/blog/the-big-data-long-tail.html.

10. Hey, T., S. Tansley, and K. Tolle, eds. *The Fourth Paradigm: Data-Intensive Scientific Discovery*. Microsoft Corporation, October 2009. http://research.microsoft.com/en-us/collaboration/fourthparadigm/.

11. Worldwide Large Hadron Collider Grid (WLCG). http://wlcg.web.cern.ch/.

12. ATLAS Experiment. http://atlas.ch/.

13. Low-Frequency Array (LOFAR). http://www.lofar.org/.

14. Google BigQuery. https://cloud.google.com/products/big-query.

15. Perry, T. S. The making of Facebook's graph search. August 6, 2013. http://spectrum.ieee.org/telecom/internet/the-making-of-facebooks-graph-search.

16. Cole, J. How Twitter stores 250 million tweets a day using MySQL. December 19, 2011. http://highscalability.com/blog/2011/12/19/how-twitter-stores-250-million-tweets-a-day-using-mysql.html.

17. Biodiversity. http://www.globalissues.org/issue/169/biodiversity.

18. Keys to Innovation—Integrated Solutions Enabling Seamless Multimodality for End Users. European Innovation Partnership for SmartCities and Communities. http://www.eu-smartcities.eu/sites/all/files/SMP%20KI%20-%20Enabling%20seamless%20multimodality%20for%20end%20users.pdf.

19. Membrey, P., K. C. C. Chan, and Y. Demchenko. A disk based stream oriented approach for storing big data. Presented at the First International Symposium on Big Data and Data Analytics in Collaboration (BDDAC 2013). Part of the 2013 International Conference on Collaboration Technologies and Systems (CTS 2013), May 20–24, 2013, San Diego, CA, USA.

20. Demchenko, Y., P. Membrey, C. Ngo, C. de Laat, and D. Gordijenko. Big security for big data: addressing security challenges for the big data infrastructure. Proceedings of the Secure Data Management (SDM'13) Workshop. Part of VLDB2013 Conference, August 26–30, 2013, Trento, Italy.

21. European Research Area. http://ec.europa.eu/research/era/index_en.htm.

22. European Union. *A Study on Authentication and Authorisation Platforms for Scientific Resources in Europe*. Final Report. Internal identification SMART-Nr 2011/0056. Brussels: European Commission, 2012. http://cordis.europa.eu/fp7/ict/e-infrastructure/docs/aaa-study-final-report.pdf.

23. *Federated Identity Management for Research Collaborations*. Final version. Reference CERN-OPEN-2012-006. https://cdsweb.cern.ch/record/1442597.

24. *SIENA European Roadmap on Grid and Cloud Standards for e-Science and Beyond*. SIENA Project report. http://www.sienainitiative.eu/Repository/Filescaricati/8ee3587a-f255-4e5c-aed4-9c2dc7b626f6.pdf.

25. Seeking new horizons: EGI's role for 2020. http://www.egi.eu/blog/2012/03/09/seeking_new_horizons_egis_role_for_2020.html.

26. UK Future Internet Strategy Group. *Future Internet Report*. May 2011. https://connect.innovateuk.org/c/document_library/get_file?folderId=861750&name=DLFE-33761.pdf.

27. *European Data Protection Directive*. http://ec.europa.eu/justice/data-protection/index_en.htm.

28. LifeWatch—e-Science European Infrastructure for Biodiversity and Ecosystem Research. http://www.lifewatch.eu/.

29. ENVRI, Common Operations of Environmental Research Infrastructure. http://envri.eu/.

30. Koopa, D., E. Santos, P. Mates, et al. A provenance-based infrastructure to support the life cycle of executable papers. *Procedia Computer Science*. 2011. http://vgc.poly.edu/~juliana/pub/vistrails-executable-paper.pdf.
31. European Commission. Open access: opportunities and challenges. European Commission for UNESCO. http://ec.europa.eu/research/science-society/document_library/pdf_06/open-access-handbook_en.pdf.
32. OpenAIR—Open Access Infrastructure for Research in Europe. http://www.openaire.eu/.
33. Open Researcher and Contributor ID. http://about.orcid.org/.
34. Data life cycle models and concepts. CEOS Version 1.2, 4 April 2012 [online] http://wgiss.ceos.org/dsig/whitepapers/Data%20Lifecycle%20Models%20and%20Concepts%20v12.docx.
35. EGI Federated Cloud Task Force. http://www.egi.eu/infrastructure/cloud/cloudtaskforce.html.
36. eduGAIN—federated access to network services and applications. http://www.edugain.org.
37. Demchenko, Y., M. Makkes, R. Strijkers, C. Ngo, and C. de Laat. Intercloud architecture framework for heterogeneous multi-provider cloud based infrastructure services provisioning. *International Journal of Next-Generation Computing (IJNGC)*, 4(2), July 2013.
38. Makkes, M., C. Ngo, Y. Demchenko, R. Strijkers, R. Meijer, and C. de Laat. Defining intercloud federation framework for multi-provider cloud services integration. Presented at the Fourth International Conference on Cloud Computing, GRIDs, and Virtualization (Cloud Computing 2013), May 27–June 1, 2013, Valencia, Spain.
39. Cloud Reference Framework. Internet draft, version 0.7. October 7, 2014 [online] http://www.ietf.org/id/draft-khasnabish-cloud-reference-framework-07.txt.
40. Turk, M. A chart of the big data ecosystem, take 2. 2012. http://mattturck.com/2012/10/15/a-chart-of-the-big-data-ecosystem-take-2/.
41. Amazon Big Data. http://aws.amazon.com/big-data/.
42. Microsoft Azure Big Data. http://www.windowsazure.com/en-us/home/scenarios/big-data/.
43. IBM Big Data Analytics. http://www-01.ibm.com/software/data/infosphere/bigdata-analytics.html.
44. Middleton, A. M. *HPCC Systems: Introduction to HPCC (High Performance Computer Cluster)*. LexisNexis Risk Solutions. May 24, 2011. http://cdn.hpccsystems.com/whitepapers/wp_introduction_HPCC.pdf.
45. Cloudera Impala Big Data Platform. http://www.cloudera.com/content/cloudera/en/home.html.
46. 10 hot big data startups to watch in 2013. January 10, 2013. http://beautifuldata.net/2013/01/10-hot-big-data-startups-to-watch-in-2013/.
47. Demchenko, Y., L. Gommans, C. de Laat, M. Steenbakkers, V. Ciaschini, and V. Venturi. VO-based dynamic security associations in collaborative grid environment. Proceedings of the 2007 International Symposium on Collaborative Technologies and Systems (CTS 2006), May 14–17, 2006, Las Vegas.
48. Demchenko, Y., C. de Laat, O. Koeroo, and D. Groep. Re-thinking grid security architecture. In *Proceedings of IEEE Fourth eScience 2008 Conference, December 7–12, 2008, Indianapolis, IN, USA*. Washington, DC: IEEE Computer Society, 2008, pp. 79–86.

49. Demchenko, Y., L. Gommans, and C. de Laat. Using SAML and XACML for complex resource provisioning in grid based applications. In Proceedings of the IEEE Workshop on Policies for Distributed Systems and Networks (POLICY 2007), Bologna, Italy, June 13–15, 2007.
50. Demchenko, Y., C. M. Cristea, and C. de Laat. XACML policy profile for multi-domain network resource provisioning and supporting authorisation infrastructure. IEEE International Symposium on Policies for Distributed Systems and Networks (POLICY 2009), July 20–22, 2009, London.
51. Research Data Alliance (RDA). http://rd-alliance.org/.
52. NIST Big Data Working Group (NBD-WG). http://bigdatawg.nist.gov/home.php.

3

Securing Cloud Data

Sushmita Ruj and Rajat Saxena

CONTENTS

Summary

Clouds are increasingly being used to store personal and sensitive information such as health records and important documents. We address the problem of storing sensitive information in the cloud so that the cloud service provider cannot tamper with the stored data. We present three problems: computing on encrypted data, access control of stored data, and auditing techniques for integrity verification. The first problem uses a cryptographic primitive called *homomorphic encryption*; the second problem uses *attribute-based encryption* (*ABE*), and the third uses *provable data possession* (*PDP*) and *proof of retrievability* (*PoR*). We survey recent results and discuss some open problems in this domain.

3.1 Introduction

Security is an important aspect of cloud computing because much information is sensitive. For example, private clouds are increasingly being used for storing medical records. There are also proposals for digitizing health records and storing them in public clouds. This not only will enable patients to access their information from anywhere in the world but also will enable other patients to seek suggestions depending on their symptoms and diseases. The patient's name and vital details can be hidden so that other patients can access their records without knowing the identity of the patient. This will benefit researchers, doctors, and other patients. Since health information is sensitive, proper measures should be taken to secure the data.

Another area of interest is social networks. The data are stored in clouds and can be accessed from anywhere using the Internet. With the growing interest in Facebook, Twitter, LinkedIn, and other social and professional networks, there is a need to protect the privacy of individuals. Privacy protection and access control are central to social networking. Security and privacy issues have been addressed [19, 21].

The following are the important security vulnerabilities in the cloud:

1. Data theft or loss: The cloud servers are distrusted in terms of both security and reliability. The cloud servers are prone to Byzantine attacks, in which they might fail in arbitrary ways. The cloud service provider (CSP) might also corrupt the data, sell data, or violate service-level agreements (SLAs). Administration errors may cause data loss during backup and restore and data migration.

2. Privacy issues: The CSP must make sure that the customer's personal information is protected from other users.

3. Infected application: Applications running on the cloud can be malicious and corrupt servers, user devices, and other applications.

4. Threats in virtualization: There are many inherent security issues in virtualization. Since clouds make extensive use of virtualization techniques, they are prone to vulnerabilities in virtualization.

5. Cross-VM (virtual machine) attack via side channels: A cross-VM attack exploits the multitenancy of the VM that enables VMs belonging to different customers to coreside on the same physical device.

Thus, the cloud should provide:

1. Availability: User data should be accessible from anywhere at any time.

2. Reliability: User data should be backed up so that even in case of failure, the data are available.

3. Integrity: Data should be available to the user as is, without any modification by the CSP or a malicious user.

4. Confidentiality: The cloud provider should not be able to read or modify data stored by the user.

5. Privacy: A user's data can be stored without knowing the actual identity of the data.

6. Accountability: The cloud should be accountable for any operation (alteration or deletion) made on the data and should not be able to refute the action.

3.1.1 Organization of the Chapter

In this chapter, we do not discuss virtualization security. We focus on secure computing using homomorphic encryption, access control using attribute-based encryption (ABE), and data auditing using provable data possession (PDP) and proofs of retrievability (PoR). For each of these security aspects, we first discuss the underlying cryptographic technique and then present how it is used to ensure cloud data security. We then present the state of the art. Section 3.2 presents homomorphic encryption for secure computation; Section 3.3 presents access control techniques using ABE. Data auditing is presented in Section 3.4. We conclude with some open problems in Section 3.5.

3.2 Homomorphic Encryption for Secure Computation in the Cloud

The cloud is being increasingly used in scientific computation. In many situations, the computation can be on sensitive data. For example, two competing companies, X and Y, have outsourced computation to the cloud.

The cloud must not be able to read the data from X and disclose the information to Y. It is thus important to hide the data from the cloud, such that the cloud operates on the encrypted data and returns the result without even knowing what data were involved. To ensure that the cloud is not able to read the data while performing computations on it, many homomorphic encryption techniques have been suggested [12, 33]. Using homomorphic encryption, the cloud receives ciphertext of the data, performs computations on the ciphertexts, and returns the encoded value of the result. The user is able to decode the result, but the cloud does not know what data were involved. In such circumstances, it must be possible for the user to verify that the cloud returned correct results.

Several encryption techniques exist that support different homomorphisms, such as multiplicative homomorphism (RSA [30]), additive homomorphism (Paillier [28], Boneh-Goh-Nissim [5]), or the recently proposed fully homomorphic scheme [12], which can support complicated functions. We give a brief description of how the Paillier homomorphic encryption technique works.

3.2.1 Paillier Homomorphic Encryption Scheme

Given two numbers M_1 and M_2, a user might want the cloud to calculate the result $M_1 + M_2$ without the cloud knowing the values of M_1 and M_2. The protocol consists of three algorithms:

1. Key generation: This algorithm generates the public keys and global parameters, given a security parameter. Let $N = p_1 p_2$, where p_1 and p_2 are primes. Choose $g \in \mathbb{Z}^*_{N^2}$, such that g has an order that is a multiple of N. Let $\lambda(N) = lcm(p_1 - 1, p_2 - 1)$, where lcm represents the least common multiple. Then, the public key is $PK = (N, g)$, and the secret key is $SK = (\lambda(N))$.

2. Encryption: Let $M \in \mathbb{Z}_N$ be a message. Select a random number $r \in Z^*_N$. The ciphertext c is given by

$$c = E(M) = g^M r^N \bmod N^2 \tag{3.1}$$

3. Decryption: To decrypt c, M can be calculated as

$$M = D(c) = \frac{L\left(c^{\lambda(N)} \bmod N^2\right)}{L\left(g^{\lambda(N)} \bmod N^2\right)} \bmod N, \tag{3.2}$$

where the L function takes input from the set $\{u < N^2 | u = 1 \bmod N\}$ and computes $L(u) = (u - 1)/N$.

Additive homomorphism is demonstrated in the following way: Suppose $c_1 = E(M_1) = g^{M_1} r_1^N$ and $c_2 = E(M_2) = g^{M_2} r_2^N$ are two ciphertexts for M_1, $M_2 \in \mathbb{Z}_N$. Then, $(c_1 c_2 \bmod N^2) = (g^{M_1+M_2} (r_1 r_2)^N \bmod N^2)$. On decryption, we have $D(c_1.c_2 \bmod N^2) = M_1 + M_2 \bmod N$. Thus, the sum of the plaintexts can be obtained from the ciphertext without the cloud knowing the values of M_1 and M_2. We note that r^N is used only to make the homomorphic computation nondeterministic; the same message can be encrypted into different ciphertexts to prevent dictionary attacks.

Boneh, Goh, and Nissim [5] proposed a scheme capable of performing multiple additions and only one multiplication at the same time. Before explaining their homomorphic encryption technique, we define *bilinear pairing*.

3.2.2 Bilinear Pairing

For bilinear pairing, let G be a cyclic group of prime order p generated by g. Let G_T be a group of order p. We can define the map e: $G \times G \to G_T$. The map satisfies the following properties:

1. $e(u^a, u^b) = e(u, v)^{ab}$ for all $u, v \in G$ and $a, b \in \mathbb{Z}_p$.
2. Nondegenerate: $e(g, g) \neq 1$.
3. e is efficiently computable.

3.2.3 Homomorphic Encryption Using Bilinear Pairings

- $Gen(\kappa) \to (pk, sk)$: Given a security parameter κ, $Gen(\kappa)$ chooses two distinct $\frac{\kappa}{2}$-bit primes, p_1 and p_2, and sets $n = p_1 p_2$. A positive integer $T < p_2$ is selected. Two multiplicative groups G, G_T of order n are selected, and a bilinear pairing e: $(G \times G) \to G_T$ is defined. Random generators $g, u \in G$ are defined and $h = u^{p_2}$ is set, such that h is a generator of the subgroup of order p_1. The public key is $pk = (n, g, h, G, G_T, e)$, and the private key is $sk = p_1$.

- $Enc(m, pk) \to c$: Given a message $m \in$ and public key pk, $Enc(pk, m)$ chooses random $r \in R$ and calculates the ciphertext

$$c = g^m h^r \bmod n$$

- $Dec(c, sk) \to m$: Given a ciphertext $c \in C$ and a private key sk, $Dec(sk, c)$ calculates

$$c' = c^p = \left(g^p\right)^m \bmod n$$

and using Pollard's lambda [38] method calculates the discrete logarithm of c' in the base g^p.

Since h is the generator of the subgroup of order p, we have $h^p = 1 \bmod n$. Thus, c' is calculated as

$$c' = c^p \bmod n$$

$$= \left(g^m h^r\right)^p \bmod n$$

$$= \left(g^m\right)^p \bmod n \tag{3.3}$$

$$= \left(g^p\right)^m \bmod n$$

The message m is bounded by T, allowing it to be recovered in time $O\left(\sqrt{T}\right)$ using Pollard's lambda method [38].

The homomorphic property of the scheme is demonstrated in the following way: Let $c_1 = g^{m_1} h^{r_1}$ and $c_2 = g^{m_2} h^{r_2}$; then,

$$c_1 c_2 = g^{m_1 + m_2} h^{r_1 + r_2} \bmod n$$

is a valid encryption of $m_1 + m_2$,

$$c_1 g^k \bmod n = g^{m_1 + k} h^{r_1} \bmod n$$

is a valid encryption of $m_1 + k$, and

$$c_1^k \bmod n = g^{km_1} h^{r_1 k} \bmod n$$

is a valid encryption of km_1. Subtraction of encrypted messages and constants can be done using $c_1 c_2^{-1} \bmod n$ and $c_1 g^{-k} \bmod n$, respectively.

Multiplication of messages is done in the following way: Let $g_1 = e(g, g)$ and $h_1 = e(g, h)$ since g generates G, $h = g^\alpha$, for some α. Given ciphertexts $c_1, c_2,$ we choose random $r \in R$; a ciphertext to compute the product $m_1 m_2$ is given by

$$e(c_1, c_2) h_1^r = e\left(g^{m_1} h^{r_1}, g^{m_2} h^{r_2}\right) h_1^r$$

$$= e\left(g^{m_1} g^{\alpha r_1}, g^{m_2} g^{\alpha r_2}\right) h_1^r$$

$$= e\left(g^{m_1 + \alpha r_1}, g^{m_2 + \alpha r_2}\right) h_1^r \tag{3.4}$$

$$= g_1^{m_1 m_2} h_1^{m_1 r_2 + m_2 r_1 + \alpha r_1 r_2 + r}$$

$$= g_1^{m_1 m_2} h_1^{r'}$$

$r' = m_1 r_2 + m_2 r_1 + \alpha r_1 r_2 + r.$

3.2.4 Fully Homomorphic Encryption

Gentry [12, 13] proposed *fully homomorphic encryption*, which is capable of evaluating any function on encrypted data. However, the schemes are impractical for implementation by cloud users since the decryption takes place at the user end. Gentry and Halevi [14] showed that even for weak security parameters, one homomorphic operation would take at least 30 seconds on a high-performance machine (and 30 minutes for the high-security parameter) [16]. Since there are many such operations, the overall time taken is too expensive for practical use in clouds.

Recently, Naehrig et al. [26] argued that fully homomorphic encryption might not be required for data privacy while computing in the cloud. Their main thesis was that only a few operations are required and a fully homomorphic property is not necessary for practical purposes. They not only proposed a *somewhat homomorphic encryption scheme* but also optimized the pairing operations to achieve the same level of security. Using their techniques, key generation runs in 250 ms and encryption takes 24 ms, whereas decryption takes 1,526 ms on a simple personal computer (PC) with an Intel Core 2 Duo processor running at 2.1 GHz, with 3 MB L2 cache and 1 GB of memory.

This technique can be used for medical data, financial purposes, and social networks, for which privacy is important. The implementation of this technique for practical purposes is still open.

3.3 Fine-Grained Access Control

We consider the following problem for which stored data can be accessed by certain groups of users and is unaccessible to other users of the network. Common examples are that of Dropbox or Google Docs: Users store files and other documents and delegate selective access to other users. Another important application is that of health care, with medical records of patients stored in the clouds, such that authorized users can access them and unauthorized users cannot. Clouds store sensitive information about patients to enable access to medical professionals, hospital staff, researchers, and policy makers. For example, a patient might want to share certain medical data with only the doctors and nurses of certain hospitals but not the hospital staff or researchers. Social networking is yet another domain where users can store and share selective information with a selective group of friends and acquaintances but not others. Assigning selective access rights to individuals is called fine-grained access control.

Access control techniques are mainly of three types: user-based access control (UBAC), role-based access control (RBAC), and attribute-based access control (ABAC). In UBAC, the access control list (ACL) contains the list of authorized users. This is not feasible in clouds where there are many users.

Sometimes, the list of users is unknown. In RBAC (introduced in [11]), users are classified based on their individual roles. Data can be accessed by users who have matching roles, which are defined by the system. For example, in the case of medical records, the personal information regarding insurance and address might be available only to the hospital staff but not to the doctors and nurses. ABAC is wider in scope; users are given attributes, and the data have an attached access policy. Only users with a valid set of attributes, satisfying the access policy, can access the data. For instance, in the example, medical records are accessed by only the neurologist or psychiatrist in only one hospital but no others. Some advantages and disadvantages of RBAC and ABAC have been discussed [22]. Most of the work in ABAC makes use of a cryptographic primitive known as the ABE.

ABAC in clouds has been studied by several researchers [e.g., 24, 31, 32, 39, 41, 42]. Some of these focused on storage of health records [e.g., 24, 41]. Using ABE, the records are encrypted under some access policy and stored in the cloud. Users are given sets of attributes and corresponding keys by a key distribution center (KDC). The keys are computed using key generation algorithms in ABE. Only when the users have a matching set of attributes can they decrypt the information stored in the cloud.

Online social networking is yet another domain where users (members) store their personal information, pictures, music, and videos and share them with selected groups of users (friends/acquaintances) or communities to which they belong. All such information is stored in clouds and given to users who satisfy matching criteria. Access control in online social networking has been studied [18]. Most of these schemes use simple ciphertext policy attribute-based encryption (CP-ABE) to achieve access control, assuming that there is only one trusted KDC.

Before we discuss how access control is achieved in clouds, we briefly talk about ABE.

3.3.1 Attribute-Based Encryption

Attribute-based encryption was proposed by Sahai and Waters [34]. In ABE, a user is given a set of attributes by an attribute authority (AA) along with a unique identity. Identity-based encryption (IBE), proposed by Shamir [37], is a public key encryption technique that eliminates the need for certification authorities and has been extensively studied. Each user in an IBE protocol has a unique identity, and the public key is the unique information about the user. IBE is a special case of ABE. There are two main variants of ABE. Key-policy ABE (KP-ABE, proposed by Goyal et al. [15]), is ABE in which the sender has attributes and encrypts data with the attributes that it has. The receiver has access policies and receives secret keys from the AA, which are constructed using the access policy. On receiving an encrypted message, the receiver can decrypt if it has matching attributes. Ciphertext-policy ABE (CP-ABE; proposed by Bethencourt et al. [4]) is the reverse of KP-ABE; the sender has the access policy built into it. The receiver has attributes, and its secret keys are constructed using the attributes it has. A receiver can decrypt messages if its set of attributes satisfies the access policy of the sender. The access policies in these protocols are monotonic access structures that have AND, OR, or general t-out-of-n threshold structures. Nonmonotonic access structures have been studied by Ostrovsky et al. [27].

We discuss the CP-ABE technique because it has been largely used for access control in clouds.

3.3.2 Ciphertext-Policy Attribute-Based Encryption

The CP-ABE consists of the following algorithms: setup, which initializes the public key PK and master secret key MK parameters; encrypt, which encrypts the message M using the public parameters PK and the access policy A and outputs a ciphertext CT; key generation, which generates the secret key SK of the users using the master secret key MK and a set of attributes S that describe the key. The decrypt algorithm takes as input the public parameters PK and a ciphertext CT, which contains an access policy A. If the set of attributes satisfies the access policy, then the decrypt algorithm returns the message M. The access policy is represented as an access tree, with attributes at the leaves and AND, OR, and t-out-of-n threshold gates at the intermediate nodes. Note that AND and OR are special cases of threshold structures where $t = n$ and $t = 1$, respectively.

The details of the protocol are as follows:

Setup: This algorithm chooses a bilinear group G of prime order p and generator g. Let $\alpha, \beta \in \mathbb{Z}_p$ be chosen at random. A hash function $H(.)$ is defined as $H: \{0, 1\}^* \rightarrow G$, which maps binary strings to elements of G. The public key is given by

$$PK = \left(G, g, h = g^\beta, e(g,g)^\alpha\right).$$

The master key is given by $MK = (\beta, g^\alpha)$.

Encrypt(PK,M,A): The encryption algorithm takes as input the message M, the public key PK, and the access policy A and returns the ciphertext CT. The algorithm chooses a polynomial P_x for each node x in the tree. The degree of the root node R is set to $k_R - 1$, where k_R is the threshold of the root. For any node, the degree of the polynomial is $d_x = k_x - 1$, where k_x is the threshold of the node. The polynomial $P_R(0) = s$, where s is randomly chosen in $\mathbb{Z}_p$. For each node x, $P_x(0) = P_{parent(x)}(index(x))$, where $parent(x)$ is the parent of x. All other coefficients of the polynomial are chosen at random from $\mathbb{Z}_p$.

Let Y be the set of leaf nodes of A. The ciphertext is given by

$$CT = \left(\mathbb{A}, C' = Me(g,g)^{\alpha s}, C = h^s, C_y = g^{P_y(0)}, C'_y = H\left(att(y)\right)^{P_y(0)} \right)$$

KeyGen(MK, S): Let S be the set of attributes of the receiver. The AA chooses $r \in \mathbb{Z}_p$ at random and $r_j \in \mathbb{Z}_p, \forall j \in S$. The secret key SK is given by

$$SK = \left(D = g^{(\alpha+r)/\beta}, \forall j \in S : D_j = g^r H\left(j\right)^{r_j}, D'_j = g^{r_j} \right)$$

Decrypt(CT, SK, S): Let $i = att(x)$ be a leaf node. If $i \in S$, then

$$DecryptNode\left(CT, SK, x\right) = \frac{e\left(D_i, C_x\right)}{e\left(D_{i'}, C'_x\right)} \tag{3.5}$$

$$= \frac{e\left(g^r H(i)^{r_i}, g^{P_x(0)}\right)}{e\left(g^{r_i}, H(i)^{P_x(0)}\right)} \tag{3.6}$$

$$= e\left(g,g\right)^{rP_x(0)} \tag{3.7}$$

If $i \notin S$, then *DecryptNode(CT, SK, x)* = NULL.

We consider the case when x is a nonleaf node; the following steps are carried out: Let S_x be the set of child nodes of x. If there are no such sets, then return NULL. Else, F_x is calculated. Lagrange's coefficient is $\Delta_{i,S}\left(x\right) = \Pi_{j \in S, j \neq i} \dfrac{x-j}{i-j}$.

$$F_x = \Pi_{z \in S_x} F_z^{\, i, S'_x(0)}, where\, i = index\left(z\right), S'_x = \left\{index\left(z\right) : z \in S_x\right\} \tag{3.8}$$

$$= \Pi_{z \in S_x} \left(e\left(g,g\right)^{rP_z(0)}\right)^{i, S'_x(0)} \tag{3.9}$$

$$= \Pi_{z \in S_x} \left(e\left(g,g\right)^{rP_{parent(z)}\left(index(z)\right)}\right)^{i, S'_x(0)} \tag{3.10}$$

$$= \Pi_{z \in S_x} e\left(g,g\right)^{rP_z(i)\, i, S'_x(0)} \tag{3.11}$$

$$= e\left(g,g\right)^{rP_x(0)} \tag{3.12}$$

The algorithm begins by calling the function on the root node R of the tree A. If the tree is satisfied by S, then

$$DecryptNode(CT, SK, r) = e(g,g)^{rP_r(0)} = e(g,g)^{rs}$$

The algorithm then calculates M as

$$C'/\left(e(C,D)/DecryptNode(CT,SK,r)\right) = C'/\left(e\left(h^s, g^{(\alpha+r)/\beta}\right)/e(g,g)^{rs}\right) = M$$

These protocols assume that the AA is honest. This is an unrealistic assumption because, in a distributed system, authorities can fail or become corrupt. To counter this problem, Chase [8] proposed a multiauthority ABE in which there are several authorities that distribute attributes and secret keys to users. The multiple-AA coordinate using a trusted authority. Chase and Chow [9] devised a multiauthority ABE protocol that required no trusted authority. However, the main problem was that a user required at least one attribute from each of the authorities, which might not be practical. Recently, Lewko and Waters [23] proposed a completely decentralized ABE by which users could have any zero or more attributes from the authorities and not require a trusted server.

We next present the distributed ABAC scheme presented in Reference 31.

3.3.3 Distributed Access Control in Clouds

Initially, for DACC the parameters of the scheme and the size of the group are decided. The size of the group is chosen to be high, for example, $2^{32} + 1$. AA A_j selects the set of attributes L_j. An owner U_u who wants to store information in the cloud chooses a set of attributes I_u that are specific to the data it wants to encrypt. These attributes may belong to different KDCs. It then decides on the access structure and converts the access tree to a linear secret sharing scheme (LSSS) matrix R using the algorithm given in Reference 31. Depending on the attributes it possesses and the keys it receives from the KDC, it encrypts and sends the data and the access matrix. Each user is given a set of attributes when the user registers for services from owners. The attributes are not given by the cloud but by the KDCs. An ssh protocol (secure shell protocol [1]) is used to securely transfer the attribute information. KDCs give secret keys to users. When a user wants to access some information, the user asks the cloud for the data record. The cloud gives an encrypted copy of the data. If a user has a valid set of attributes, then the user calculates the data using the secret key that it possesses.

Encryption proceeds in two steps. The Boolean access tree is first converted to an LSSS matrix. In the second step, the message is encrypted and sent to the cloud along with the LSSS matrix. A secure channel like ssh can be used for the transmission.

We consider the example from Reference 31 of a network in which owners want to store their data in encrypted form in the cloud and give selective access to users. In a health care scenario, owners can be the patients who store their records in the cloud, and doctors, nurses, researchers, and insurance companies can retrieve them. There are attribute authorities, which are servers scattered in different countries, that generate secret keys for the users. AAs can be government organizations that give different credentials to users. These servers can be maintained by separate companies, so that they do not collude with each other. This differs from the concept of a cloud. A particular cloud is maintained by one company; thus, if authorities are a part of the cloud, then they can collude and find the secret keys of all the users. Figure 3.1 shows the overall model of our cloud environment. The users and owners are denoted by n_i; the AAs are servers that distribute attributes and secret keys SK to users and owners. AAs are not part of the cloud. The owner encrypts a message and stores the ciphertext C in the cloud.

Suppose an owner U_u wants to store a record M. U_u defines the access structure A, which helps it to decide the authorized set of users who can access the record M. It then creates an $m \times h$ matrix R (m is the number of attributes in the access structure) and defines a mapping function π of its rows with the attributes. π is a permutation, such that π:$\{1, 2, \ldots , m\} \rightarrow$ W. The encryption algorithm takes as input the data M that need to be encrypted, the group G, the LSSS matrix R, and the permutation function π, which maps the attributes

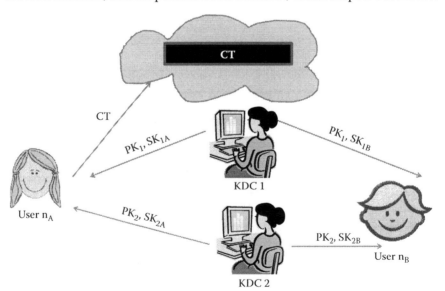

FIGURE 3.1
Distributed access control in clouds.

in the LSSS to the actual set of attributes. For each message M, the ciphertext C is calculated per Equations (3.16) and (3.17). Ciphertext C is then stored in the cloud.

When a user U_u requests a ciphertext from the cloud, the cloud transfers the requested ciphertext C using the ssh protocol. The decryption algorithm returns plaintext message M if the user has a valid set of attributes.

System initialization: Select a prime p, generator g of G, groups G and G_T of order p, a map $e: G \times G \rightarrow G_T$, and a hash function $H: \{0, 1\}^* \rightarrow G$ that maps the identities of users to G. Each AA $A_j \in A$ has a set of attributes L_j. The attributes disjoint ($L_i \cap L_j = \varphi$ for $i \neq j$). Each AA also chooses two random exponents $\alpha_i, y_i \in \mathbb{Z}_q$. The secret key of AA A_j is

$$SK[j] = \{\alpha_i, y_i, i \in L_j\}. \tag{3.13}$$

The public key of AA A_j is published:

$$PK[j] = \{e(g,g)^{\alpha_i}, g^{y_i}, i \in L_j\}. \tag{3.14}$$

Key generation and distribution by KDCs: User U_u receives a set of attributes $I[j, u]$ from AA A_j and corresponding secret key $sk_{i,u}$ for each $i \in I[j, u]$:

$$sk_{i,u} = g^{\alpha_i} H(u)^{y_i}, \tag{3.15}$$

where $\alpha_i, y_i \in SK[j]$. Note that all keys are delivered to the user securely using the user's public key, such that only that user can decrypt it using its secret key.

Encryption by sender: The sender decides about the access tree and encrypts message M as follows:

1. Choose a random seed $s \in \mathbb{Z}_q$ and a random vector $v \in \mathbb{Z}_q^h$, with s as its first entry; h is the number of leaves in the access tree (equal to the number of rows in the corresponding matrix R).
2. Calculate $\lambda_x = R_x \cdot v$, where R_x is a row of R.
3. Choose a random vector $w \in \mathbb{Z}_q^h$ with 0 as the first entry.
4. Calculate $\omega_x = R_x \cdot w$
5. For each row R_x of R, choose a random $\rho_x \in \mathbb{Z}_q$.
6. The following parameters are calculated:

$$C_0 = Me(g,g)^s$$

$$C_{1,x} = e(g,g)^{\lambda_x} e(g,g)^{\alpha_{\pi(x)}\rho_x}, \forall x$$

$$C_{2,x} = g^{\rho_x} \forall x \qquad\qquad\qquad (3.16)$$

$$C_{3,x} = g^{y_{\pi(x)}\rho_x} g^{\omega_x} \forall x,$$

where $\pi(x)$ is mapping from R_x to the attribute i that is located at the corresponding leaf of the access tree.

7. The ciphertext C is sent by the sender (it also includes the access tree via R matrix):

$$C = \langle R, \pi, C_0, \{C_{1,x}, C_{2,x}, C_{3,x}, \forall x\} \rangle \qquad\qquad (3.17)$$

Decryption by receiver: Receiver U_u takes as input ciphertext C, secret keys $\{sk_{i,u}\}$, group G, and outputs message M. It obtains the access matrix R and maps π from C. It then executes the following steps:

1. U_u calculates the set of attributes $\{\pi(x): x \in X\} \cap I_i$ that are common to itself and the access matrix. X is the set of rows of R.

2. For each of these attributes, it checks if there is a subset X' of rows of R, such that the vector $(1, 0, \dots, 0)$ is their linear combination. If not, decryption is impossible. If yes, it calculates constants $c_x \in \mathbb{Z}_q$, such that $\sum_{x \in X'} c_x R_x = (1, 0, \dots, 0)$.

3. Decryption proceeds as follows:

    (a) For each $x \in X'$, $dec(x) = \dfrac{C_{1,x} e\left(H(u), C_{3,x}\right)}{e\left(sk_{\pi(x),u}, C_{2,x}\right)}$

    (b) U_u computes $M = C_0 / \Pi_{x \in X'} dec(x)$.

None of the above techniques can authenticate users or protect the privacy of the user. It is just not enough to store the contents securely in the cloud; it might also be necessary to ensure the anonymity of the user. However, the user should be able to prove to the other users that he or she is a valid user who stored the information without revealing the identity. For example, a user would like to store some sensitive information but does not want to be recognized. The user's privacy needs to be protected when the user needs to store confidential information but does not reveal his or her identity. For example, if a user wants to store a controversial record about the employers, then he or she might want to remain anonymous. The cloud, on the other hand, must be able to authenticate the user as an authorized person. Ruj et al. [32] proposed an authentication mechanism that also protects the privacy of the user. Users cannot just read from already stored data but can be given the right to modify the data. Attribute-based signatures (ABSs) [25] are used for this purpose. In ABSs, users have a claim predicate associated with a message. The claim predicate helps to identify the user as an authorized one without revealing the user's identity. Other users or the cloud can verify the user and the validity of the message stored. An ABS can be combined with ABE to achieve authenticated access control without disclosing the identity of the user to the cloud (see Table 3.1).

Attribute-based encryption involves expensive operations, which might be burdensome on resource-constrained devices like smartphones and the like. To address this problem, Green et al. [16] proposed a technique to outsource the decryption to a proxy, such that the operations performed by the user can be done efficiently and the complex computations are delegated to the proxy. The proxy, however, cannot decrypt the information.

We present a comparison in Table 3.1 of access control schemes used in the literature. Some schemes are centralized (have a single KDC), and some are decentralized (have multiple KDCs). We look for the type of operations supported, that is, x-Write-y-Reads (denoted x-W-y-R). Some schemes have authentication and some do not. Only the Green et al. [16] scheme outsources decryption. We also check if revocation of users is permitted or not.

3.4 Data Auditing

A big challenge is to ensure that the integrity of the data is preserved. Cloud servers are prone to Byzantine failure, in which they can fail in arbitrary ways. Generally, the cloud protects data integrity by making redundant copies of data. To reduce storage space, the CSP might not offer the same degree of redundancy as presented in the SLA. The CSP might also discard rarely used data, without informing the client, just to save storage space. Thus, data auditing is needed to verify that the cloud has not tampered with the stored data.

Data auditing is mostly done in a probabilistic way, in which a few blocks are chosen and verified. The commonly used techniques are as follows:

1. Provable data possession (PDP): Allows the client to verify that the cloud has stored the original data faithfully without retrieving it.
2. Proofs of retrievability (PoR): The cloud should be able to prove that it has stored the client's data correctly, and the client is able to extract the data from the cloud.

We note that the difference between PDP and PoR techniques is that PDP techniques only produce a proof for recoverable data possession, but PoR

TABLE 3.1

Comparison of Our Scheme with Existing Access Control Schemes

Scheme Reference No.	Centralized/ Decentralized	Write/Read Access	Type of Access Control	Privacy-Preserving Authentication	Decryption Outsourcing?	User Revocation?
24	Centralized	1-W-M-R	ABE	No authentication	No	No
41	Centralized	1-W-M-R	ABE	No authentication	No	No
31	Decentralized	1-W-M-R	ABE	No authentication	No	Yes
16	Centralized	1-W-M-R	ABE	No authentication	No	No
42	Centralized	M-W-M-R	ABE	Authentication	No	No
32	Decentralized	M-W-M-R	ABE	Authentication	No	Yes
16	Centralized	1-W-M-R	ABE	No Authentication	Yes	Yes

schemes check the possession of data and can recover data in case of data access failure or data loss. Usually, a PDP scheme can be transformed into a PoR scheme by adding erasure or error-correcting codes.

The early definitions of PoR [20] and PDP [2] used the definitions in a general client server setting; however, we define it in the context of the cloud. We discuss each of these models and present some third-party (public) auditing techniques in which anyone can verify the data that a client has stored.

3.4.1 Provable Data Possession Techniques

The PDP schemes involve a challenge/response protocol between the client (verifier) and the CSP (prover). It consists of two main steps:

- The client (verifier) first allows the CSP (prover) to store files.
- Later, the client can verify if the CSP possesses the data by challenging the CSP.

The PDP techniques generate probabilistic proofs of possession by sampling random sets of blocks from the server; this drastically reduces input/output (I/O) costs. In PDP techniques, the client maintains a constant amount of metadata to verify the proof. The challenge/response protocol transmits a low, constant amount of data that minimize network communication. Thus, the PDP schemes for remote data checking support large data sets in widely distributed storage systems.

Ateniese et al. [2] were the first to define PDP schemes formally. Later, they [3] proposed a very lightweight and provable secure data possession scheme in the random oracle model. This scheme is based entirely on symmetric key cryptography. The main idea of this scheme is that, before outsourcing, a client precomputes a certain number of short possession verification tokens, each token covering some set of data blocks. The actual data are then handed over to the CSP. Subsequently, when the client wants to obtain a proof of data possession, the client challenges the data storage server with a set of random block indices. In turn, the data server must compute a short integrity check over the specified blocks (corresponding to the indices) and return it to the client. For the proof to hold, the returned integrity check must match the corresponding value precomputed by the client. However, in their scheme, the client has the choice of either keeping the precomputed tokens locally or outsourcing them in encrypted form to the server. In the latter case, the client's storage overhead is constant regardless of the size of the outsourced data. The scheme is also efficient in terms of storage, computation overheads, dynamic support for data operations, and bandwidth. Sebé et al. [35] presented a scheme that used asymmetric key cryptography (RSA modules) for integrity verification.

Erway et al. [10] presented a fully dynamic provable data possession (DPDP), which extends the PDP model to support provable updates to stored

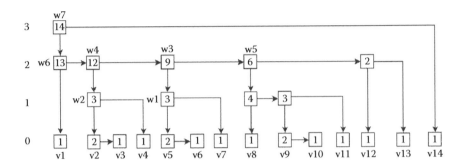

FIGURE 3.2
Example of rank-based skip list.

data. They used rank-based authenticated skip lists (Figure 3.2) and aggregate signatures [6]. Before discussing the scheme, we discuss authenticated skip lists and aggregate signatures.

Skip lists [29] are a probabilistic alternative to balanced trees. Balancing a data structure probabilistically is easier than explicitly maintaining the balance and is easy to implement. Skip lists are also space efficient. In a skip list, each node v stores two pointers, denoted $rgt(v)$ and $dwn(v)$, that are used for searching. $l(v)$ is the level of the node v; $l = 0$ denotes the leaf nodes. An authenticated skip list that uses a collision-resistant hash function can be used to check the integrity of file blocks.

3.4.2 Rank-Based Skip Lists

Let F be a file consisting of n blocks $m_1, m_2, \ldots, m_n$. At the ith bottom-level node of the skip list, the signature $x(m_i)$ of block m_i is stored. Block m_i is stored separately at the cloud. Each node v of the skip list stores the number of nodes at the bottom level that can be reached from v. This value is called the *rank* of v and is denoted by $r(v)$.

The top leftmost node of a skip list is referred to as the start node. For a node v, $low(v)$ and $high(v)$ denote the indices of the leftmost and rightmost nodes at the bottom level reachable from v, respectively. Clearly, for the start node S of the skip list, $r(S) = n$, $low(S) = 1$, and $high(S) = n$. Using the ranks stored at the nodes, the ith node of the bottom level can be reached by traversing a path that begins at the start node as follows: For the current node v, assume that $low(v)$ and $high(v)$ are known.

Let $w = rgt(v)$ and $z = dwn(v)$. The following values are set:

$$high(w) = high(v), \quad low(w) = high(v) - r(w) + 1$$

$$high(z) = low(v) + r(z) - 1, \quad low(z) = low(v)$$

To reach the ith bottom node, we start from $v = S$, where S is the start node while (ith bottom node is not reached):

```
{
    if i ∈ [low(w), high(w)]
        v = w is set   //the right pointer is followed
    else
        v = z is set   //the down pointer is followed
}
```

For each node v of the rank-based authenticated skip list, a label $f(v)$ is defined as follows:

$$f(v) = \begin{cases} 0 & \text{if } v = null \\ h\big(l(v), r(v), f\big(dwn(v)\big), f\big(rgt(v)\big)\big) & \text{if } l(v) > 0 \\ h\big(l(v), r(v), x(v), f\big(rgt(v)\big)\big) & \text{if } l(v) = 0 \end{cases}$$

The next two algorithms have been described [10], but we present them here for completeness.

Algorithm 1 Path Generation:
$$PathGen(i) \rightarrow \{x(v_i), \Pi\}$$

```
1: Let v₁, ..., vₖ be the verification path for block i
2:    return signature x(vᵢ) of block i and the table
      Π = (A(v₁), ..., A(vₖ)) corresponding to block i
```

Algorithm 2 Skip List Verification:
$$verify(\Pi, \ x(v_i), \ sig_{sk}(H(f(S)))) \rightarrow \{TRUE, \ FALSE\}$$

```
1:  Let Π = (A₁, ..., Aₖ),
2:  where Aⱼ = (dⱼ, lⱼ, qⱼ, gⱼ),1 ≤ j ≤ k
3:  λ₁ = 0; ρ₁ = 1 + q₁; δ₁ = d₁; ζ₁ = 0;
4:  γ₁ = h(λ₁, ρ₁, x(vᵢ), g₁);
5:  for j = 2, ..., k do
6:      λⱼ = lⱼ; ρⱼ = ρⱼ₋₁ + qⱼ; δⱼ = dⱼ;
7:      if δⱼ = = rgt then
8:          γⱼ = h(λⱼ, ρⱼ, gⱼ, γⱼ₋₁);
9:          ζⱼ = ζⱼ₋₁;
10:     else if δⱼ = = dwn then
11:         γⱼ = h(λⱼ, ρⱼ, γⱼ₋₁, gⱼ);
12:         ζⱼ = ζⱼ₋₁ + qⱼ;
13      end if
14: end for
```

```
15: if e(sig_sk(H(f(S))),g) ≠ e(H(γ_k),v) then
16:    return FALSE
17: else if ρ_k - ζ_k ≠ i then
18:    return FALSE
19: else
20:    return TRUE
21: end if
```

3.4.3 Skip List Verification

First, we describe the algorithm *PathGen(i)* [10] to generate a verification path for block *i*. The verification path is the reverse search path; for example, let $v_k, v_{k-1}, \ldots, v_1$ be the search path for block *i*, then $v_1, v_2, \ldots, v_k$ is the verification path for block *i*. For each node $v_j, j = 1, \ldots, k$, Boolean $d(v_j)$ and values $q(v_j)$ and $g(v_j)$ are defined as follows, where *r(null)* is set to 0:

$$
d(v_j) = \begin{cases} rgt & j = 1 \, or \, j > 1 \, and \, v_{j-1} = rgt(v_j) \\ dwn & j > 1 \, and \, v_{j-1} = dwn(v_j) \end{cases}
$$

$$
q(v_j) = \begin{cases} r(rgt(v_j)) & j = 1 \\ 1 & j > 1 \, and \, l(v_j) = 0 \\ r(dwn(v_j)) & j > 1, l(v_j) > 0 \, and \, d(v_j) = rgt \\ r(rgt(v_j)) & j > 1, l(v_j) > 0 \, and \, d(v_j) = dwn \end{cases}
$$

$$
g(v_j) = \begin{cases} f(rgt(v_j)) & j = 1 \\ x(v_j) & j > 1 \, and \, l(v_j) = 0 \\ f(dwn(v_j)) & j > 1, l(v_j) > 0 \, and \, d(v_j) = rgt \\ f(rgt(v_j)) & j > 1, l(v_j) > 0 \, and \, d(v_j) = dwn \end{cases}
$$

The *PathGen(i)* algorithm returns the sequence $\Pi(i) = (A(v_1), \ldots, A(v_k))$ where $A(v) = (d(v), l(v), q(v), g(v))$ for the block *i* with signature *x(i)*. Table 3.2 shows the sequence $\Pi(v_6)$ as the sample verification path. Due to the properties of the skip list, data in the verification path have an expected size $O(\log n)$ with high probability.

To verify the skip list, the verifier requests the signature generated by the client by signing the start node of the skip list and the path table Π for any random block. Then, the verifier runs the skip list verification algorithm

TABLE 3.2

PathGen Table for the Sixth Block of the File F Stored
in the Skip List in Figure 3.2

Node v	v_6	v_5	w_1	w_3	w_4	w_6	w_7
$D(v)$	rgt	rgt	dwn	dwn	rgt	rgt	dwn
$l(v)$	0	0	1	2	2	2	3
$Q(v)$	0	1	1	6	3	1	1
$G(v)$	0	$x(v_5)$	$f(v_7)$	$f(w_5)$	$f(w_2)$	$f(v_1)$	$f(v_{14})$

(Algorithm 2) to verify the integrity of the skip list using the table Π and the signature of the start node S, which is $sig_{sk}(H(f(S)))$, sent by the cloud server.

Algorithm 2 iteratively computes tuples $(\lambda_j, \rho_j, \delta_j, \gamma_j)$ for each node v_j on the verification path plus a sequence of integers ζ_j. At each iteration of the for-loop, the tuple $(\lambda_j, \rho_j, \delta_j, \gamma_j)$ associated with a node v_j of the verification path represents the following:

- $\lambda_j = l(v_j)$, that is, the level of v_j;
- $\rho_j = r(v_j)$, that is, the rank of v_j;
- δ_j indicates whether we arrived at v_j from right or below;
- $\gamma_j = f(v_j)$, that is, the label of v_j;
- ζ_j is equal to the sum of the ranks of all the nodes that are to the right of nodes of the path seen so far but are not on the verification path.

3.4.4 Bilinear Aggregate Signatures

We have used BLS (Boneh, Lynn, and Shacham) aggregate signatures in our scheme to achieve public auditability and blockless verification.

3.4.4.1 BLS Signature Scheme

Boneh, Lynn, and Shacham [6] gave a simple, deterministic signature scheme in which the signatures are very short. The signer's secret key is $x \in \mathbb{Z}_q$, the public key is $y = g^x$, and g is the generator of the multiplicative group G of order q. Let $H : \{0, 1\}^* \to G$ be a hash function. The signature scheme is given by

Sign(m): the signature σ on message m is $\sigma = (H(m))^x \in G$.

Verify(σ, m): accept if $e(g, \sigma) = e(g^x, H(m))$.

3.4.4.2 Aggregate Signature Scheme

Aggregate signatures [6] are used if we have different signers who want to sign different messages but we only want to produce one signature. This is useful for instances such as batch auditing or certificate chains. The signer i has secret key x_i and public key $y_i = g^{x_i}$ and wants to sign message m_i; we assume that all messages are distinct.

Sign$(m_1, ..., m_n)$: First, each signer computes its signature: $\sigma_i = H\left(m_i\right)^{x_i}$, $1 \leq i \leq n$. The aggregate signature $\sigma = \prod\limits_{i=1}^{n} \sigma_i$.

Verify(m, σ): Accept if $e\left(g, \sigma\right) = \prod\limits_{i=1}^{n} e\left(y_i, H\left(m_i\right)\right)$.

3.4.5 Data Auditing Using Aggregate Signatures

We present the basic algorithms used in the data-auditing protocol.

KeyGen$(1^k) \rightarrow (pk, sk)$: This probabilistic algorithm is run by the client. It takes a security parameter 1^k and returns public key pk and secret key sk.

SigGen$(sk, F) \rightarrow (\Phi, sig_{sk}(H(f(S))))$: This algorithm is run by the client. It takes as input private key sk and *file F*, which is an ordered collection of blocks m_i, and outputs a signature set $\Phi = \{\sigma_i\}_{i = 1,2,...,n}$. It also outputs metadata: the signature $sig_{sk}(H(f(S)))$ of the start node S of a rank-based authenticated skip list. In our construction, the level zero nodes of the rank-based authenticated skip list contain hashes $H(m_i)$.

SSig$_{ssk}(\cdot)$: It is a signing function that uses signing key ssk to sign a string.

GenProof$(F, \Phi, \Psi) \rightarrow (P)$: This algorithm is run by the server. It takes as input a file F, its signatures Φ, and a challenge Ψ (discussed further in the chapter). It outputs a data integrity proof P for the blocks specified by the challenge Ψ.

VerifyProof$(pk, \Psi, P) \rightarrow \{TRUE, FALSE\}$: This algorithm can be run by a verifier on the receipt of P. It takes as input public key pk, the challenge Ψ, and proof P returned by the server and outputs *TRUE* if the integrity of the file is verified as correct and *FALSE* otherwise.

ExecUpdate$(F, \Phi, update) \rightarrow (F', \Phi', P_{update})$: This algorithm is run by the server. It takes as input a file F, its signatures Φ, and a data operation request "update" from the client. It outputs updated file F', updated signatures Φ', and a proof P_{update} for the operation.

VerifyUpdate$(pk, sig_{sk}(H(f(S))), update, P_{update}) \rightarrow \{(TRUE, FALSE, sig_{sk}(H(S')))\}$: This algorithm is run by the client. It takes as input public key pk, the signature $sig_{sk}(H(f(S)))$, operation request "update," and the proof P_{update} from the server. If verification succeeds, it outputs a signature $sig_{sk}(H(S'))$ for the new start node S' or *FALSE* otherwise.

3.4.6 Third-Party Auditing of Cloud Data

We assume that file F (potentially encoded using Reed-Solomon codes) is divided into n blocks $m_1, m_2, \ldots, m_n$, where $m_i \in \mathbb{Z}_q$ and q is a large prime. Let $e: G \times G \rightarrow G_T$ be a bilinear map and $H: \{0, 1\}^* \rightarrow G$ be a hash function that converts binary strings to elements of G and is viewed as a random oracle. Let g be the generator of G.

The data-auditing scheme consists of the following steps:

- *Setup*: This step initializes the system and generates public and secret keys.

  1. The client generates a random signing key pair (ssk, spk) by invoking $KeyGen(1^k)$. Then, the client chooses a random $\alpha \in \mathbb{Z}_q$ and computes $v = g^\alpha$. The secret key is $sk = \{\alpha, ssk\}$, and the public key is $pk = \{v, spk\}$.

  2. $SigGen(\cdot)$ is invoked to preprocess the file F and to generate metadata before sending the file to the cloud server. Given $F = (m_1, m_2, \ldots, m_n)$, the client chooses a random element $u \in G$. $SigGen(sk, F)$ is invoked to preprocess the file F and to generate metadata before sending the file to the cloud server. Let $t = file\ name||n||u||SSig_{ssk}(filename||n||u)$ be the file tag for F. Then, the client computes signature σ_i for each block $m_i(i = 1, 2, \ldots, n)$ as $\sigma_i = \left(H(m_i).u^{m_i} \right)^\alpha$. We denote the set of signatures by $\Phi = \{\sigma_i\}_{1 \leq i \leq n}$.

  3. The client generates the rank-based skip list, where the bottom level nodes contain the hashes of m_i, $1 \leq i \leq n$ denoted by $H(m_i)$.

  4. The client signs the hash $H(f(S))$, where S is the start node and $f(S)$ is the label of the start node. The client signs using the private key α: $sig_{sk}(H(f(S))) \leftarrow (H(f(S)))^\alpha$.

  5. The client sends $\{F, t, \Phi, sig_{sk}(H(f(S)))\}$ to the cloud server.

  6. The client now deletes $\{F, \Phi, sig_{sk}(H(f(S)))\}$.

- *Integrity verification protocol*: Once the client has stored the data on the cloud storage server, the verification protocol can be initiated. The client can also perform the integrity verification on the data using a similar process or the task can be delegated to a third-party auditor (TPA).

  1. The TPA first uses spk to verify the signature on t. If the verification fails, TPA returns *FALSE*; otherwise, it recovers u from t.

  2. The TPA chooses a random value $r \in [1, n]$ and requests the cloud server to send the table $\Pi(r)$. The cloud server runs Algorithm 1 to calculate $\Pi(r)$.

3. After receiving $\Pi(r)$, the TPA runs Algorithm 2 to verify the skip list stored on the cloud server and retrieves $f(S)$, where $f(S)$ is the label of the start node.

4. Now, the TPA determines a suitable c (the number of blocks to be verified) according to the desired probability of error detection and ω.

5. The TPA generates a challenge Ψ, picks a random c-element subset $I = \{s_1, s_2, \ldots, s_c\}$ of set $[1, n]$, where we assume $s_1 \leq \ldots \leq s_c$. Then, a random element $v_i \subseteq \mathbb{Z}_p$. The challenge Ψ specifies the positions of the blocks to be checked. The TPA sends $\Psi = \{(i, v_i)\}_{s_1 \leq i \leq s_c}$ to the prover (server).

6. After this, the server generates the proof P for each of the challenges sent by the TPA. On receiving a challenge $\Psi = \{(i, v_i)\}_{s_1 \leq i \leq s_c}$, the server computes $\mu = \sum\limits_{i=s_1}^{s_c} v_i m_i \in \mathbb{Z}_q$ and $\sigma = \prod\limits_{i=s_1}^{s_c} \sigma_i^{v_i} \in G$, where both the data blocks and the corresponding signature blocks are aggregated into a single block, respectively. The server also sends the signatures of the requested blocks as the set $\{\Omega_i\}_{s_1 \leq i \leq s_c}$. The server then sends the proof $P = \{\mu, \sigma, \{\Omega_i\}_{s_1 \leq i \leq s_c}\}$ to the TPA.

7. After receiving the proof P to the corresponding challenge Ψ, the TPA verifies the integrity by checking

$$e(\sigma, g) \overset{?}{=} e\left(\prod_{i=s_1}^{s_c} H(m_i)^{v_i} \cdot u^\mu, v \right)$$

If this equation holds, then TPA returns *TRUE*, *FALSE*, otherwise.

A similar technique is used during an update.

Zhu et al. [43] (MULTI-PDP or multiple PDP) addressed the construction of an efficient PDP scheme for distributed cloud storage to support the dynamic scalability of service and data migration. For this, they consider the coexistence of multiple CSPs to cooperatively store and maintain the client's data. This scheme is based on a homomorphic verifiable response and hash index hierarchy. Security for this scheme is based on a multiprover zero-knowledge proof system, which can satisfy knowledge soundness, zero-knowledge, and completeness properties.

3.4.7 Proof-of-Retrievability Schemes

The idea of PoR schemes is to verify a small number of blocks, chosen at random, instead of the whole file. If there are errors, then the file can be retrieved using error-correcting codes.

The scheme of Reference 20 uses *sentinels*, which are check blocks and are randomly embedded in the file. The file is then encrypted, such that it is impossible to detect the positions of the sentinels. The client sends a set of sentinel positions and asks the CSP to return the value of the sentinels. If the CSP has modified or deleted the data, then, with a high probability, it is impossible to return the values of the sentinels. In this approach, encryption renders the sentinels indistinguishable from other file blocks.

If the number of sentinels queried is small, it might not be possible to correctly detect the errors, but with the error-correcting codes, it is possible to recover the file. If the number of sentinels queried is large, then the user might not be able to retrieve the file correctly but will be able to detect that tampering of the file has occurred.

The scheme has six basic functions [20]: The function "Respond" is the only function executed by CSP P. All other functions are executed by the verifier (client) V. The set of verifier-executed functions modifies some persistent state α. π presents the complete collection of system parameters.

1. Keygen $[\pi] \to \kappa$: The function Keygen generates a secret key κ.
2. Encode $(F, \kappa, \alpha) \to (\tilde{F}_\eta, \eta)$: The function Encode generates a file handle η that is unique to a given verifier invocation. The function also transforms file F into an (enlarged) file $\tilde{F}_\eta$ and provides the pair $(\tilde{F}_\eta, \eta)$ as an output.
3. Extract(η, κ, α)$[\pi] \to F$: It determines a sequence of challenges that V sends to P and processes the resulting responses. If successful, the function recovers and outputs F_η.
4. Challenge(η, κ, α)$[\pi] \to c$: Challenge takes the secret key κ and a handle η and accompanying state α as input, along with system parameters. The function outputs a challenge value c for the file η.
5. Respond(c, $\eta \to r$): The function Respond is used by P to generate a response to a challenge c. This challenge can originate with either the Challenge or the Extract function.
6. Verify($(r, \eta)\kappa, \alpha \to b \in (0,1)$: The function verify determines whether r is a valid response to challenge c. The function outputs a "1" bit if verification succeeds and "0" otherwise.

A basic unit of storage is an l-bit block. The error-correcting code operates over l-bit symbols, a cipher operates on l-bit blocks, and the sentinels have l bits. The file consists of b blocks (b is a multiple of k) and is $F = (f_1, f_2, \dots, f_b)$. The function Encode consists of the following steps:

1. Error correction: To each k blocks an (n, k, d)-error-correcting code is applied, and the resulting file is $F' = (f_1, f_2, \dots, f_{b'})$, with $b' = bn/k$.
2. Encryption: A symmetric cipher is used, and F' is converted to F". The cipher is so chosen that each block can be separately decrypted.

3. Sentinel created: s Sentinels are created and appended to F'' to yield F'''.

4. Permutation: The $b' + s$ blocks of the file F''' are permuted to yield $\tilde{F}$.

The prover (CSP) produces a concise proof that the archive retains and reliably transmits the entire file or data object F. To ensure that the archive has retained F, the verifier (client) V challenges the prover by specifying the positions of a collection of sentinels in $\tilde{F}$ and asking to return the associated sentinel values. This phase includes Extract, Challenge, Respond, and Verify functions. If the sentinels are returned correctly, then the file has not been tampered with; if there are errors, then the error-correcting code is used to retrieve the message. A drawback of this PoR scheme is the preprocessing/encoding of F required prior to storage with the prover.

Shacham et al. [36] utilized two new economic and efficient homomorphic authenticators. These authenticators are the primary encryption or hashing. They also need larger storage requirements on the prover and provides proof of security against impulsive adversaries.

Bowers et al. [7] introduced HAIL (high-availability and integrity layer), a general conceptual framework for PoRs that is an improvement [20, 36]. It claims lower storage requirements and a higher level of security assurance with minimal computational overhead and tolerates higher error rates than scheme [20]. It is robust against an active, mobile adversary, that is, one that may progressively corrupt the full set of servers. This work describes design challenges encountered for practical implementation of PoR protocols. HAIL is a distributed cryptanalytic system that allows a set of servers to prove to a client that a stored file is intact and retrievable. Building blocks of the HAIL system are the universal hash function, message authentication codes (MACs), and integrity-protected error-correcting codes (IP-ECC). The advantage of the HAIL adversary security model is that it ensures distributed file system availability against a strong, mobile adversary.

The drawbacks of the PoR and PDP schemes are as follows:

- The effectiveness of these schemes rests primarily on the preprocessing steps that the user conducts before outsourcing the data file. This introduces significant computation and communication complexity.
- Most of these techniques do not support privacy preservation and dynamic data operations.
- Most of these schemes focus on only static and archive data.
- None of these schemes considers batch auditing.

Public verifiability is needed in many cases when others should be able to verify the data. A trusted TPA might have expertise and technical capabilities that the clients do not have. Data audits by a trusted third party (TTP) involve an independent authenticated entity to conduct a data audit.

Wang et al. [40] determined the difficulties and potential security issues of direct extensions for fully dynamic data updates and then constructed

TABLE 3.3

Comparison of Data-Auditing Schemes

Scheme	Public Verifiability	Data Dynamics	Privacy Preserving	Detection Probability
PDP				
Ateniese et al. [2]	Yes	Append only	No	$1-(1-p)^c$
Ateniese et al. [3]	No	Yes	No	$1-(1-p)^c$
Sebé et al. [35]	No	No	No	$1-(1-p)^c$
Erway et al. [10]	No	Yes	No	$1-(1-p)^c$
PoR				
Juels and Kaliski [20]	No	No	No	$1-(1-p)^c$
Shacham and Waters [36]	No	No	No	$1-(1-p)^c$
HAIL: Bowers et al. [7]	No	No	No	$1-(1-p)^c$
Wang et al. [40]	Yes	Yes	Yes	$1-(1-p)^{cs}$
Hao et al. [17]	Yes	Yes	Yes	$1-(1-p)^{cs}$

a verification scheme that takes these issues into account. Specifically, to achieve efficient data dynamics, they improved the present proof of storage models by manipulating the classic Merkle hash tree construction for block tag authentication. They explored a bilinear aggregate signature to support efficient handling of multiple auditing tasks and extend output into a multiuser setting, where TPA can perform multiple auditing tasks simultaneously. This theme achieves batch auditing wherever multiple delegated auditing tasks from totally different users are often performed at the same time by the TPA.

Hao et al. [17] described a remote data integrity-checking protocol that supports public verifiability, data dynamics, and privacy against verifiers without any TPA. They used RSA-based homomorphic verifiable tags for their protocol construction.

Table 3.3 shows the comparative analysis of different data-auditing schemes. We indicate if the scheme is probabilistic or deterministic, whether public verifiability is satisfied, if the scheme can support dynamic data, and if the scheme is privacy preserving. We compare the detection probability in each case. Here, c is the number of blocks sampled, and p is the probability that a block is corrupted.

3.5 Conclusion and Future Work

In this chapter, we discussed a few security issues in cloud computing. Most of the techniques help us to protect against dishonest CSPs. Cloud security also involves other aspects, for example, virtualization security, not addressed here.

There are many security challenges that need to be addressed.

3.5.1 Security in Mobile Clouds

Most of the cryptographic techniques are computation intensive. This might not be a good option for mobile devices, which are energy constrained. So, efficient encryption and decryption protocols need to be devised to enable security on mobile devices. One way is to outsource some of the encryption and decryption operations to a third party or a proxy server.

3.5.2 Distributed Data Auditing for Clouds

In most of the related work on data auditing, the auditor is assumed to be a trusted party. However, this is a strong assumption. Thus, distributed auditing looks attractive. This will make the auditing process more robust. Assigning all auditing jobs to one TTP can also slow the whole system. Thus, a distributed auditing service not only will balance the load but also will provide trustworthy service. Users with idle resources can contribute toward distributed data auditing.

3.5.3 Secure Multiparty Computation on Clouds

Secure multiparty computation is a cryptographic paradigm in which n users compute a function securely, keeping their inputs private. The users send their inputs in such a way that only the function can be computed, without knowing the individual input. These computations are extremely involved and are good candidates for computation on clouds. However, a single server is prone to single-point failure. For this reason, distributed computing on clouds is an attractive option. Secure multiparty computation in clouds is a promising area of research.

References

1. Secure shell protocol. http://tools.ietf.org/html/rfc4252.
2. Giuseppe Ateniese, Randal C. Burns, Reza Curtmola, Joseph Herring, Lea Kissner, Zachary N. J. Peterson, and Dawn Xiaodong Song. Provable data possession at untrusted stores. In Peng Ning, Sabrina De Capitani di Vimercati, and Paul F. Syverson, editors, *ACM Conference on Computer and Communications Security*, pages 598–609. New York: ACM, 2007.
3. Giuseppe Ateniese, Roberto Di Pietro, Luigi V. Mancini, and Gene Tsudik. Scalable and efficient provable data possession. *IACR Cryptology ePrint Archive*, 2008:114, 2008.

4. John Bethencourt, Amit Sahai, and Brent Waters. Ciphertext-policy attribute-based encryption. In *IEEE Symposium on Security and Privacy*, pages 321–334. Washington, DC: IEEE Computer Society, 2007.

5. Dan Boneh, Eu-Jin Goh, and Kobbi Nissim. Evaluating 2-dnf formulas on ciphertexts. In Joe Kilian, editor, *TCC*, volume 3378 of *Lecture Notes in Computer Science*, pages 325–341. New York: Springer, 2005.

6. Dan Boneh, Ben Lynn, and Hovav Shacham. Short signatures from the Weil pairing. In Colin Boyd, editor, *ASIACRYPT*, volume 2248 of *Lecture Notes in Computer Science*, pages 514–532. New York: Springer, 2001.

7. Kevin D. Bowers, Ari Juels, and Alina Oprea. HAIL: a high-availability and integrity layer for cloud storage. *IACR Cryptology ePrint Archive*, 2008:489, 2008.

8. Melissa Chase. Multi-authority attribute based encryption. In *TCC*, volume 4392 of *Lecture Notes in Computer Science*, pages 515–534. New York: Springer, 2007.

9. Melissa Chase and Sherman S. M. Chow. Improving privacy and security in multi-authority attribute-based encryption. In Ehab Al-Shaer, Somesh Jha, and Angelos D. Keromytis, editors, *ACM Conference on Computer and Communications Security*, pages 121–130. New York: ACM, 2009.

10. C. Christopher Erway, Alptekin Küpçü, Charalampos Papamanthou, and Roberto Tamassia. Dynamic provable data possession. In Ehab Al-Shaer, Somesh Jha, and Angelos D. Keromytis, editors, *ACM Conference on Computer and Communications Security*, pages 213–222. New York: ACM, 2009.

11. David F. Ferraiolo and D. Richard Kuhn. Role-based access controls. In *15th National Computer Security Conference*, 1992.

12. Craig Gentry. Fully homomorphic encryption using ideal lattices. In Michael Mitzenmacher, editor, *STOC*, pages 169–178. New York: ACM, 2009.

13. Craig Gentry. Toward basing fully homomorphic encryption on worst-case hardness. In Tal Rabin, editor, *CRYPTO*, volume 6223 of *Lecture Notes in Computer Science*, pages 116–137. New York: Springer, 2010.

14. Craig Gentry and Shai Halevi. Implementing gentry's fully-homomorphic encryption scheme. In Kenneth G. Paterson, editor, *EUROCRYPT*, volume 6632 of *Lecture Notes in Computer Science*, pages 129–148. New York: Springer, 2011.

15. Vipul Goyal, Omkant Pandey, Amit Sahai, and Brent Waters. Attribute-based encryption for fine-grained access control of encrypted data. In Ari Juals, Rebecca Wright, and Sabrina De Capitani di Vimercati, editors, *ACM Conference on Computer and Communications Security*, pages 89–98. New York: ACM, 2006.

16. Matthew Green, Susan Hohenberger, and Brent Waters. Outsourcing the decryption of abe ciphertexts. In David Wagner, editor, *USENIX Security Symposium*. Berkeley, CA: USENIX Association, 2011.

17. Zhuo Hao, Sheng Zhong, and Nenghai Yu. A privacy-preserving remote data integrity checking protocol with data dynamics and public verifiability. *IEEE Trans. Knowl. Data Eng.*, 23(9):1432–1437, 2011.

18. Sonia Jahid, Prateek Mittal, and Nikita Borisov. Easier: encryption-based access control in social networks with efficient revocation. In Bruce S. N. Cheung, Lucas Chi Kwong Hui, Ravi S. Sandhu, and Duncan S. Wong, editors, *ASIACCS*, pages 411–415. New York: ACM, 2011.

19. Wayne Jansen and Timothy Grance. *Guidelines on Security and Privacy in Public Cloud Computing*. NIST Special Publication 800-144. Gaithersburg, MD: NIST, 2011.

20. Ari Juels and Burton S. Kaliski Jr. Pors: proofs of retrievability for large files. In Peng Ning, Sabrina De Capitani di Vimercati, and Paul F. Syverson, editors, *ACM Conference on Computer and Communications Security*, pages 584–597. New York: ACM, 2007.

21. Seny Kamara and Kristin Lauter. Cryptographic cloud storage. In Radu Sion, Reza Curtmola, Sven Dietrich, Aggelos Kiayias, Josep M. Miret, Kazue Sako, and Francesc Sebé, editors, *Financial Cryptography Workshops*, volume 6054 of *Lecture Notes in Computer Science*, pages 136–149. New York: Springer, 2010.

22. D. Richard Kuhn, Edward J. Coyne, and Timothy R. Weil. Adding attributes to role-based access control. *IEEE Computer*, 43(6):79–81, 2010.

23. Allison B. Lewko and Brent Waters. Decentralizing attribute-based encryption. In Kenneth G. Paterson, editor, *EUROCRYPT*, volume 6632 of *Lecture Notes in Computer Science*, pages 568–588. New York: Springer, 2011.

24. Ming Li, Shucheng Yu, Kui Ren, and Wenjing Lou. Securing personal health records in cloud computing: patient-centric and fine-grained data access control in multi-owner settings. In Sushil Jajodia and Jianying Zhou, editors, *SecureComm*, pages 89–106, Singapore, 2010.

25. Hemanta K. Maji, Manoj Prabhakaran, and Mike Rosulek. Attribute-based signatures: achieving attribute-privacy and collusion-resistance. *IACR Cryptology ePrint Archive*, 2008:328, 2008.

26. Michael Naehrig, Kristin Lauter, and Vinod Vaikuntanathan. Can homomorphic encryption be practical? In Christian Cachin and Thomas Ristenpart, editors, *CCSW*, pages 113–124. New York: ACM, 2011.

27. Rafail Ostrovsky, Amit Sahai, and Brent Waters. Attribute-based encryption with non-monotonic access structures. In Peng Ning, Sabrina De Capitani di Vimercati, and Paul F. Syverson, editors, *ACM Conference on Computer and Communications Security*, pages 195–203. New York: ACM, 2007.

28. Pascal Paillier. Public-key cryptosystems based on composite degree residuosity classes. In Jacques Stern, editor, *EUROCRYPT*, volume 1592 of *Lecture Notes in Computer Science*, pages 223–238. New York: Springer, 1999.

29. W. Pugh. Skip lists: a probablistic alternative to balanced trees. *Commun. ACM*, 33(6):668–676, 1990.

30. R. L. Rivest, L. Adleman, and M. L. Dertouzos. On data banks and privacy homomorphisms. In Richard Lipton, David Dobkin, and Anita Jones, editors, *Foundations of Secure Computation*, Orlando, FL, 1978.

31. Sushmita Ruj, Amiya Nayak, and Ivan Stojmenovic. DACC: distributed access control in clouds. In Huaimin Wang, Stephen R. Tate, and Yang Xiang, editors, *Proceedings of IEEE TrustCom*, pages 91–98, Changsha, China, 2011.

32. Sushmita Ruj, Milos Stojmenovic, and Amiya Nayak. Privacy preserving access control with authentication for securing data in clouds. In *CCGRID*, pages 556–563. New York: IEEE, 2012.

33. Ahmad-Reza Sadeghi, Thomas Schneider, and Marcel Winandy. Token-based cloud computing. In Alessandro Acquisti, Sean W. Smith, and Ahmad-Reza Sadeghi, editors, *TRUST*, volume 6101 of *Lecture Notes in Computer Science*, pages 417–429. New York: Springer, 2010.

34. Amit Sahai and Brent Waters. Fuzzy identity-based encryption. In Ronald Cramer, editor, *EUROCRYPT*, volume 3494 of *Lecture Notes in Computer Science*, pages 457–473. New York: Springer, 2005.

35. Francesc Sebé, Josep Domingo-Ferrer, Antoni Martínez-Ballesté, Yves Deswarte, and Jean-Jacques Quisquater. Efficient remote data possession checking in critical information infrastructures. *IEEE Trans. Knowl. Data Eng.*, 20(8):1034–1038, 2008.
36. Hovav Shacham and Brent Waters. Compact proofs of retrievability. *IACR Cryptol. ePrint Arch.*, 2008:73, 2008.
37. Adi Shamir. Identity-based cryptosystems and signature schemes. In *CRYPTO*, pages 47–53, 1984.
38. Douglas Stinson. *Cryptography: Theory and Practice.* Boca Raton, FL: CRC Press, 2005.
39. Guojun Wang, Qin Liu, and Jie Wu. Hierarchical attribute-based encryption for fine-grained access control in cloud storage services. In Ehab Al-Shaer, Angelos D. Keromytis, and Vitaly Shmatikov, editors, *ACM Conference on Computer and Communications Security*, pages 735–737. New York: ACM, 2010.
40. Qian Wang, Cong Wang, Kui Ren, Wenjing Lou, and Jin Li. Enabling public auditability and data dynamics for storage security in cloud computing. *IEEE Trans. Parallel Distrib. Syst.*, 22(5):847–859, 2011.
41. Shucheng Yu, Cong Wang, Kui Ren, and Wenjing Lou. Attribute based data sharing with attribute revocation. In Bruce S. N. Cheung, Lucas Chitturi, Ravi Sandhu, and Duncan Wong, editors, *ACM ASIACCS*, pages 261–270, Hong Kong: ACM, 2010.
42. Fangming Zhao, Takashi Nishide, and Kouichi Sakurai. Realizing fine-grained and flexible access control to outsourced data with attribute-based cryptosystems. In Feng Bao and Jian Weng, editors, *ISPEC*, volume 6672 of *Lecture Notes in Computer Science*, pages 83–97. New York: Springer, 2011.
43. Yan Zhu, Hongxin Hu, Gail-Joon Ahn, and Mengyang Yu. Cooperative provable data possession for integrity verification in multicloud storage. *IEEE Trans. Parallel Distrib. Syst.*, 23(12):2231–2244, 2012.

4

Adaptive Execution of Scientific Workflow Applications on Clouds

Rodrigo N. Calheiros, Henry Kasim, Terence Hung, Xiaorong Li, Sifei Lu, Long Wang, Henry Palit, Gary Lee, Tuan Ngo, and Rajkumar Buyya

CONTENTS

Summary

Many e-science applications can be modeled as workflow applications. In this programming model, scientific applications are described as a set of tasks that have dependencies between them. Clouds are natural candidates for hosting such applications. This is because some of their core characteristics, such as rapid elasticity, resource pooling, and pay per use, are well suited to the nature of scientific applications that experience variable demand, spikes in resource (i.e., of the central processing unit [CPU] or disk) utilization, and sometimes, urgency for generation of results. As current workflow management systems (WfMSs) cannot support efficient and automated execution of workflow in clouds that support adaptive execution, fault tolerance, and data privacy, in this chapter we detail the requirements of a WfMS that supports these requirements, its architecture, and an application scenario involving simulation of Singapore's public transport system.

4.1 Introduction

Many e-science applications can be modeled as *workflow applications*. In this programming model, scientific applications are described as a set of tasks that have dependencies between them. Normally, this dependency is expressed in the form of input and output (I/O) files. It means that, before one task can execute, it needs the tasks it depends on to have completed their execution and the files they generate to already be available as input. Well-known application domains where workflow applications are used include astrophysics, bioinformatics, and disaster modeling and prediction, among others.

Scientists have been successfully executing this type of application on supercomputers, clusters, and grids. Recently, with the advent of clouds, scientists started investigating the suitability of this infrastructure for workflow applications.

Clouds are natural candidates for hosting workflow applications. This is because some of their core characteristics, such as rapid elasticity, resource pooling, and pay per use, are well suited to the nature of scientific applications that experience variable demand, spikes in resource (i.e., of the central processing unit [CPU], disk) utilization, and sometimes, urgency for generation of results. Furthermore, recent offerings of high-performance cloud computing instances make it even more compelling for scientists to adopt clouds as the platform of choice for hosting their scientific workflow applications.

The execution of workflow applications is a demanding task. Tasks, sometimes in the order of hundreds, need to have their execution coordinated. They have to be submitted for execution in a specific virtual machine (VM), and the required input files need to be made accessible for the application. This may require the transfer of huge amounts of data between computing hosts. Reception of user input, data transfers, task executions, and VMs can fail; in this case, some action has to be carried out to reestablish the execution of the application. Examples of such actions are retrying the data transfer, rescheduling the task, or starting a new VM to execute the remaining tasks. These activities are carried out by software called *workflow management systems* (WfMSs). Examples of well-know, WfMSs are Pegasus [1], Taverna [2], Triana [3], and Cloudbus Workflow Engine [4].

At the same pace that infrastructures and platforms evolve, so do the scientific applications using such infrastructures and platforms. The amount of data generated by scientific experiments is reaching the order of terabytes per day, and huge capacity is required to process this data to enable scientific discoveries. Therefore, WfMSs also need to evolve to support huge data sets and the complex analytics required to extract useful insights from the generated data. Even more important, if data are continuously generated, WfMSs need to support real-time capabilities. This has to be achieved at the same time that other nonfunctional requirements, such as data privacy, are enabled.

Although this information is truth regardless of the specific infrastructure hosting the workflow application, even more complexity is added to the system when the applications are executed in clouds. This is because extra capabilities are required to enable the WfMS to select the right number of resources of the right type so that the computational task is performed within a user-defined time frame and budget.

As current WfMSs cannot support efficient and automated execution of workflow in clouds that support adaptive execution, fault tolerance, and data privacy, we developed extensions to a workflow engine [4] to support such features. In this chapter, we detail the requirements of such a system, its architecture, and the application scenario explored, along with an evaluation of the system and a discussion of lessons learned during its development.

4.2 Workflow Applications

The workflow programming model is undoubtedly one of the most prominent programming models in e-science, being used in a range of domains, including bioinformatics, astrophysics, and disaster modeling, to name a few. In this model, one application (job) is composed of a number of tasks that have execution dependencies between them. Typically, the dependency is related to I/O: One task depends on the output of another (or other) task(s) as its input; therefore, it cannot be executed until such data are available (normally, after the execution of the original task is completed).

Variations of the model exist in which the workflow also contains conditional branches (i.e., particular tasks that compose the workflow may or may not be executed depending on the results of previous tasks), loops (for which execution of specific sections of the workflow is repeated), and when tasks are allowed to start execution before predecessors complete execution.

Without loss in generality, a workflow application can be formally represented by a directed acyclic graph (DAG) whose vertices represent tasks and the directed edges represent the dependencies between tasks: An edge A → B indicates that task B depends on task A for its execution. Such a representation of workflow applications is also known as DAG. A simple workflow is depicted in Figure 4.1.

Traditionally, workflow applications have been extensively deployed in high-performance infrastructures such as supercomputers and clusters [5]. When deployed on such infrastructures, emphasis was given in reducing the execution time of the workflow by optimizing the utilization of the resources available for the workflow. When grids became available, they were also used for workflow execution [6, 7]. This added complexity to the scheduling process because it was possible that resources available for execution

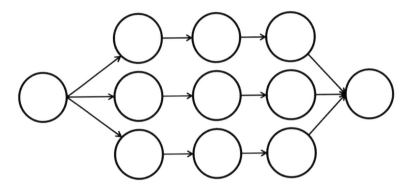

FIGURE 4.1
Graphical representation of a simple scientific workflow.

were distributed, and thus data movement across wide distances might be necessary. Even in this case, focus was still on execution time minimization.

Cloud computing adds a new dimension for workflow execution related to the financial cost of using a virtually infinite amount of resources for workflow execution. This means that the only limitations to the available resources, and consequently the improvements in execution time, are the available *budget* for workflow execution and the *structure of the workflow* itself, which determines the maximum amount of tasks that can be executed in parallel in the infrastructure. Clouds also brought other challenges for workflow management and execution. They are discussed in the next section.

4.3 Requirements for Adaptive Execution of Workflows on Clouds

Although modern WfMSs already support clouds as the platform supporting the execution of workflow applications, many desirable features are still absent in the WfMSs. This is because current WfMSs for clouds are derived from projects in the area of grid computing. Therefore, many of their features are optimized for grids and thus are unable to obtain the most key aspects of clouds, such as rapid elasticity.

In this sense, clouds add extra complexity to WfMSs because the amount of resources that WfMSs can provision for executing the workflow is virtually infinite, as long as there is budget available to spend on the workflow execution process. Thus, different from existing algorithms and approaches that operated with the goal of obtaining the most from the resources available for the application, cloud-enabled WfMSs can assume that the main restriction of the system is the budget rather than resources, and its goal is balancing utilization, cost, and reduction of execution time [8].

Li et al. [8] also identified the following requirements for cloud-enabled workflows:

1. *Dynamic resource provisioning and deadlines:* This is the capability of acquiring and releasing resources as required to accommodate the tasks of the workflow and to enable their completion within a user-specified deadline. This is an important feature because it enables execution of *mission-critical* workflow applications that need to be completed before the deadline for the computation to have value. An example of such mission-critical workflows is disaster management workflows. Consider, for example, the architecture depicted in Figure 4.2. A disaster management workflow application suite may support management of many types of natural disasters, such as floods, cyclones, and bushfires. When one such disaster strikes, the corresponding management application needs to be executed in public and private clouds to provide information that will be used by disaster mitigation and rescue teams. If the application takes too long to execute, the teams will not have time to act based on the information provided, which results in wasted time (and money) invested in the execution of the workflow in the cloud and even further losses in terms of lives and property damage that would have been prevented if the rescue and mitigation teams had access to the information in appropriate time.

2. *Adaptive task/workflow/user scheduling:* This relates to the capability of reacting to conditions faced during workflow execution to maintain the balance between cost, utilization, and execution time. In the context of this requirement, a change in conditions means adapting to changes in user requirements at runtime (e.g., increased/reduced budget, increased/reduced application deadline).

3. *Fault tolerance:* This is the capacity to automatically react to changes in the available number of resources or tasks to be processed because of failures and the capability to adapt to situations if the performance delivered by cloud resources is below that contracted or historically observed.

4. *Security-conscious data migration and data privacy:* Given that the data being processed by the WfMS can be sensitive, mechanisms for protection of the data, either during transfer or once stored in a public cloud, must be available. The applied method should also enable auditing of accesses and modifications in the data.

5. *Application management:* This requirement involves the capability to collect and process information about the system status and monitor the platform and the application in real time. This requirement also includes a capacity for presentation of comprehensive information to users about the resources (utilization, performance, etc.) and

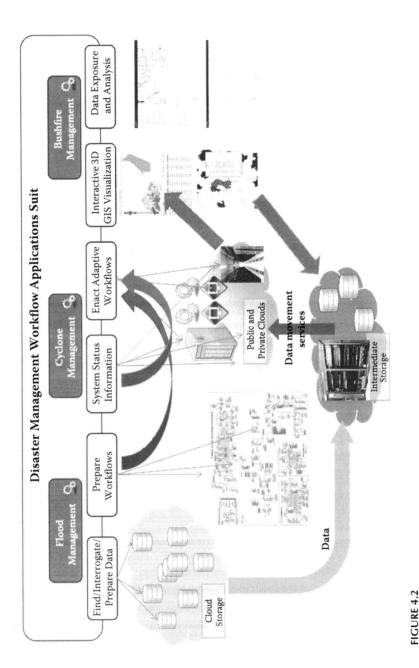

FIGURE 4.2
Architecture for workflow-enabled disaster management applications.

tasks so the cost-benefit analysis of utilization of the cloud can be undertaken and the utilization of cloud computing for workflows can be justified.

These requirements were addressed while we developed an adaptive system for execution of a workflow for agent-based simulation in hybrid clouds. The application is detailed next.

4.4 Case Study

A city is sustainable only if it can accommodate economic and population growth while ensuring the well-being of its people and environment [9]. Therefore, reaching sustainability becomes harder when the growth of a population is high or when the growth occurs in areas of high density, such as Singapore.

Singapore's land area has increased from 581 km^2 in the 1960s to 716 km^2 in 2012; its population in the same period has grown from 1.6 million to 5.3 million [10]. To maintain reasonably good economic growth, the Singapore government has projected a need for the population to reach 6.9 million by 2030. However, the land area is only slated to grow to about 800 km^2 in the same period. The disparity in the growth rate of population versus land area means that there is increasing strain on space and the service infrastructure. It is crucial for the planning agencies to adopt a scientific approach to understanding the urban fabric and how it can adapt to social, economic, and environmental changes.

One key aspect to improve the quality of living of city inhabitants is public transport. There is a need for efficient transport covering the biggest extension of the city as possible and running with enough frequency so people are motivated to use it rather than using cars. In this sense, Singapore's public transport network is ranked among the best in the world. Its Mass Rapid Transit (MRT) train network comprises 102 stations distributed over four main lines, with a total of almost 150 km of rail lines. It currently serves around 2.5 million commuters per day, which represents more than 75% of the total public transport users [11].

The number of commuters and the high frequency of trains (running in intervals as short as 90 seconds) make it a complex system. Furthermore, even a minimal disruption in the operation of one train can cascade over several lines, affecting hundreds of thousands of commuters.

This complex and sensitive system will be subject to even further pressure as the population increases. Therefore, tools are needed to help planners evaluate the effects that disruptions would have over the whole system.

This fact motivated us to adopt a data-driven approach to understanding the dynamics of the public transport system in Singapore. To achieve that, a scalable complex system modeling for a sustainable city (S^3) has been developed to study how the city will behave under different planning scenarios.

The goal of S^3 is to provide insights to users on what-if scenarios for a day-to-day public transport system by leveraging on a synthetic journey function that generates agent-based models for public transport dynamics simulation. This insight will provide information on the future public transport infrastructure preparedness to handle the growing population and the preparedness for emergencies in cases of breakdowns in the public transport system.

Scaling areas that we address in this context are (1) the extract-transform-load (ETL) or preprocessing that is required to train the synthetic journey function that generates the agent-based model; (2) the agent-based generation required to generate millions of agents that represent the increasing population and public transport infrastructure; and (3) the large-scale agent-based simulation that is required to handle, track, and process each of the agents and to support complex interactions between agents to provide insight on what-if scenarios for the public transportation system in Singapore.

We tackled the large-scale computation requirements by designing agent-based complex system modeling supported by an adaptive cloud WfMS [12] for workflow scheduling and handling big data and dynamic resource scaling on public and private clouds.

The S^3 application has three phases: preprocessing, data analysis, and agent-based simulation. Figure 4.3 shows our S^3 application architecture, which comprises an adaptive cloud WfMS, ETL or preprocessing algorithm, data analysis algorithm, and agent-based simulation.

> **ETL or preprocessing**. The synthetic data set for the application is based on the studies of trends and random sampling of daily public commuters' activities in Singapore. It consists of 1-second time granularities for 7 days' duration with approximately 3 million journeys per day. Based on the synthetic data set, we extract and transform the data for travel duration for each origin-station to destinations-station (OD-pair) of 90 x 90 by three different route choices. The order of complexity in this phase is $O(n^2)$, where n represents the number of stations.

> **Data analysis**. The objective of this phase is to understand commuter demand and, based on data analysis results, create or improve the journey function of all possible OD pairs, possible routes for each OD pair, and temporal travel demand. The order of complexity in this phase is also $O(n^2)$, where n represents the number of stations.

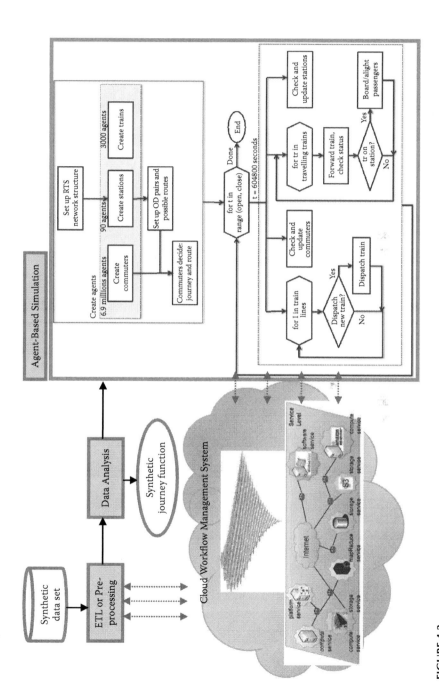

FIGURE 4.3
S³: architecture, concepts, and technologies.

Agent-based simulation. In this phase, we simulate the actions and interactions of autonomous agents. This agent-based simulation consists of agent granularity, adaptive agent process, decision-making heuristics, and agent interactions. Agent granularity refers to the number of agents specified at various scales. The adaptive agents process refers to the action that an agent takes when a situation occurs (redefining the decision-making heuristics). Decision-making heuristics refer to rules or behaviors of an agent. Agent interaction refers to the complexity of communications or interactions between agents.

There are three types of agents in the S^3 application: commuters, stations, and trains. Each of these agents has its own attributes, adaptive agent process, decision-making process, and agent interactions, as summarized in Table 4.1. The order of complexity in this phase is $O(n^3)$ due to the interactions between agents on simulation time interval or $O(tn^2)$, where n represents the number of agents and t represents the simulation time steps.

Data requirements. The size and quantity of the data set that is generated is large. The size of the data can easily take up a few gigabytes each day. For example, the data set consists of 7 days of public transportation journeys for each individual, with approximately 3 million journeys per day. As for the agent-based simulation, we simulate the growing population as 6.9 million. This translates into approximately 14 million journeys (travel and return) performed for each simulated day.

Computation requirements. For agent-based simulation, millions of agents are created to simulate the future infrastructure and dynamics of the transportation system in Singapore. In total, the system manages 7 million agents that have their own attributes, adaptive agent process, decision-making process, and interactions with other agents. Furthermore, there is complexity of agent interactions and tracking for the simulation interval at 1-second granularity.

TABLE 4.1

Agent-Based Simulation Characteristics

	Commuter Agents	**Station Agents**	**Train Agents**
Agent granularity	6.9 million agents	90 agents	Approximately 200 agents
Attributes	12 attributes	9 attributes	16 attributes
Adaptive agent process	1 adaptive process	—	2 adaptive processes
Decision-making heuristics	5 decision-making heuristics	2 decision-making heuristics	5 decision-making heuristics
Agent interactions	• Station • Train	• Commuter • Train	• Commuter • Station

To support not only these requirements for data and computation but also the requirements listed in the previous section, we proposed and developed a workflow middleware whose architecture is described next.

4.5 System Architecture

The requirements presented previously are addressed by software middleware comprising a WfMS augmented with capabilities for data analytics integrated as a second layer above the WfMS. The overall organization of the system is depicted in Figure 4.3. It shows the S^3 application architecture, which consists of the adaptive cloud WfMS, the ETL or preprocessing algorithm, the data analysis algorithm, and the agent-based simulation.

Cloud WfMS system. The cloud WfMS is responsible for workflow scheduling, big data handling, and dynamic resource scaling on hybrid clouds. The Cloud WfMS comprises the workflow engine, task dispatcher, and resource management. The workflow scheduling coordinates the execution of tasks, handles communication between components, implements the scheduling algorithm, and manages the execution of applications on distributed resources. The task dispatcher component submits tasks to resources for execution. The resource management component interacts with the cloud infrastructure to enable resource allocation.

Preprocessing and data analysis. This component is responsible for managing preprocessing and data analysis activities that are required to train the synthetic journey function that generates the synthetic journey. It tackles the scalability challenge by dynamically scaling up the number of VM instances; thus, the preprocessing processes are executed in parallel. Since this is a computationally intensive task with a long duration and the total number of origin-station and destination-station pairs is large (composed of more than 8,000 pairs), VM instances are pooled from a hybrid cloud where each VM instance processes the travel duration for each origin-station and destination-station pair.

Agent-based simulation. There are three phases of agent-based simulation: agent creation, attribute definition, and simulation execution. Our module is able to scale the process of agent-based generation in orders of magnitude of up to millions of agents. Further in this chapter, we demonstrate the process for 6.9 million commuter agents, 90 station agents, and 200 train agents. The activities of the process of simulation execution are (1) time series simulation with 1-second intervals; (2) tracking of each agent, which includes checking and updating each agent's state; (3) a decision-making process for each agent (e.g., dispatch the train at simulation time *t*); (4) adaptive agent process that allows agents to adapt to different situations (e.g., when a train arrives at a station, commuter agents need to board or leave the train); (5) interactions between agents (e.g., communication between train agents and station agents when the train arrives at the station, communication between commuter agents and train agents when the commuter boards the train) and management of tasks and data flows on the hybrid cloud utilizing the cloud WfMS.

A discussion of the implementation aspects of the architecture and its performance is presented next.

4.6 Discussion and Lessons Learned

The agent-based simulation is based on three phases: create agents, define attributes, and run the simulation. To test the scalability of the model, we evaluated two different setups. The first one uses the ZeroMQ (ZMQ) technology [13] in our hybrid cloud. ZMQ is a low-latency asynchronous message-passing library that is used in scalable distributed or concurrent applications. The second one is a hybrid cloud test bed. The private cloud component of the hybrid cloud is composed of 64 cores (hyperthreaded) and a 2.2-GHz processor with 128 GB of memory. On top of this infrastructure, we deployed 50 VMs, with each VM an Ubuntu 12.04 with 1 core and 4 GB of memory. The public cloud is composed of 1,000 Amazon EC2 small instances (1 core with 1 ECU and 1.7 GB of memory).

Scaling of the "create agents" and "define attributes" phases is achieved through the division of the workload, with each process handling a group of agents. For example, in a simulation with 7 million commuters running on an infrastructure containing 1,000 VMs, creation of commuter agents was split among the VMs in such a way that each VM handled the creation of 7,000 agents.

On the "run simulation" phase, we experienced the execution of the simulation on a time-based simulation with 1-second intervals and tracking, checking, and updating of each agent's states. The scale method in this case delegates each VM to handle a group of agents. When the ZMQ push-pull method is used, one of the VMs acts as the head node that is in charge of distributing the tasks to all the worker VMs and controlling the timekeeping process of the simulation. The timekeeping process consists of sending a message to each worker to inform them of the current simulation time so that workers can start the simulation of events scheduled for such a given time.

However, we noticed that the time-based simulation has limited scalability. When executed in a private cloud of 50 VMs, it took 35,248 seconds to complete the 2 million commuters' agent-based simulation. This happened because there were dependencies in $t + 1$ with time t (i.e., simulation at time t needs to be completed before simulation of time $t + 1$ starts). Because of this issue, we replaced the time-based simulation with an event-based simulation.

In event-based simulation, the model handles the agents' interactions, such as boarding of commuters, unboarding of commuters, train arrivals at stations, and train departures from stations. On the back end, the workload is distributed via a similar method to other phases (each process handles a group of agents). With this new technique, the execution time of the simulation in the same private cloud was completed in 1,818 seconds for the same 2-million-commuter agent-based simulation, an improvement of 19 times over the original technique.

We further scaled the agent-based simulation by executing it on 1,000 VMs. In this case, the agent-based simulation completed in 434 seconds for simulation of 2 million commuters and 963 seconds for 7 million commuters. This demonstrated that the three phases of our approach are scalable and suitable for execution on elastic cloud platforms.

To summarize, we gave preference to the cloud-enabled WfMS over the ZMQ system because of the following reasons: (1) It enabled more efficient management of the highly distributed data required by the agent-based simulation workflow; (2) it better automated the workflow process for data analytics with multiobjective optimization of performance and budget; and (3) it enabled dynamic resource allocation for adaptive services with fault tolerance.

4.7 Related Work

Given the importance of workflow applications for the scientific community, many scientific workflow platforms were developed to explore scientific computational platforms such as grids. As cloud platforms became popular among the scientific community, WfMSs where enhanced to support them.

Pegasus [1] offers a set of tools for different aspects of execution and management of workflow applications and platforms. It implements application programming interfaces (APIs) for diverse programming languages, supports submission of workflows via web portals, and integrates with external tools. On its back end, it supports multiple cloud providers and scientific infrastructures.

Taverna [2] is another widely adopted workflow engine that can explore both grid and cloud platforms. Applications running on the platform can be deployed in many modes, including "server mode," by which it supports requests from many users to execute remote workflow applications.

The Cloudbus Workflow Engine incorporates a market-oriented utility computing model that supports grids, desktops, and clouds. It supports the concept of InterCloud for allocation and management of resources for execution of workflow applications [1].

Kim et al. [14] proposed a WfMS able to deploy workflows in hybrid infrastructures composed of TeraGrid nodes and Amazon EC2 resources. Our proposed system, on the other hand, can also leverage resources from private and public cloud providers.

Gogouvitis et al. [15] proposed a WfMS for deploying workflow applications on virtualized environments that is able to utilize resources from public clouds. However, it has no dynamic provisioning capabilities to speed application execution and to meet real-time application performance requirements as does our approach.

Fernandez et al. [16] proposed a cloud WfMS that applies a concept called chemical programming for the application scheduling. The system, however, does not offer dynamic resource provisioning capabilities and autonomic self-healing features.

CometCloud [17] is a more recent tool that implements an infrastructure for autonomic management of workflow applications on clouds.

4.8 Conclusions and Future Work

Clouds became a powerful platform for e-research as they enable scientists to have access to elastic, cost-effective, and virtually infinite computing power. Because clouds provide their users the view of infinite computing capacity, the real limitations on the scalability of the applications lie in the available budget for cloud usage and limitations in the applications themselves. Therefore, it is important that scientific application developers enable their applications to get the most from the cloud.

In this chapter, we discussed recent trends for execution of workflows in clouds. The architecture we presented is composed of a platform layer and an application layer. The platform layer enables operations such as dynamic resource provisioning, autonomic scheduling of applications, fault tolerance, security, and privacy in data access. The features enabled by this layer can be explored by virtually any application that can be described as scientific workflow.

In the application layer, we discussed a data analytics application enabling simulation of the public transport system of Singapore and the effect of abnormal events in the transport network. The application consists of an agent-based simulation of the public transport system of Singapore, and it allows evaluation of effects of incidents (such as train delays) in the flow of passengers in the country.

As future work, we plan to extend our platform to support a disaster decision support system (DDSS). The principles presented in this chapter will be further expanded so the DDSS will provide a dashboard for the strategic, tactical, and operational decisions arising during disaster mitigation. It will be integrated with a range of modeling and simulation tools to provide optimization models with up-to-date situational awareness and predictions to provide recommendations to authorities. This extension will support not only workflow applications but also other programming models suitable for clouds, such as MapReduce. Ideally, the platform will support not only applications that are entirely described as one of these models but also complex applications that are composed of diverse subcomponents that may be developed as different programming models.

References

1. Deelman, E., Singh, G., Su, M., et al. 2005. Pegasus: a framework for mapping complex scientific workflows onto distributed systems. *Scientific Computing* 13:219–237.
2. Oinn, T., Greenwood, M., Addis, M., et al. 2006. Taverna: lessons in creating a workflow environment for the life sciences. *Concurrency and Computation: Practice and Experience* 18:1067–1100.
3. Taylor, I., Shields, M., Wang, I., et al. 2007. The Triana Workflow Environment: Architecture and Applications. In *Workflows for E-Science*, ed. I. J. Taylor, E. Deelman, D. B. Gannon, et al., 320–339. London: Springer.
4. Pandey, S., Karunamoorthy, D., and Buyya, R. 2011. Workflow engine for clouds. In *Cloud Computing: Principles and Paradigms*, ed. R. Buyya, J. Broberg, and A. Goscinski, 321–344. New York: Wiley.
5. Kwok, Y., and Ahmad, I. 1999. Static scheduling algorithms for allocating directed task graphs to multiprocessors. *ACM Computing Surveys* 3:406–471.
6. Yu, J., Buyya, R., and Ramamohanarao, K. 2008. Workflow scheduling algorithms for grid computing. In *Metaheuristics for Scheduling in Distributed Computing Environments*, ed. F. Xhafa and A. Abraham, 173–214. Berlin: Springer.
7. Hirales-Carbajal, A., Tchernykh, A., Yahyapour, R., et al. 2012. Multiple workflow scheduling strategies with user run time estimates on a grid. *Journal of Grid Computing* 10:325–346.
8. Li, X., Calheiros, R., Lu, S., et al. 2012. Design and development of an adaptive workflow-enabled spatial-temporal analytics framework. In *Proceedings of the 2012 IEEE International Workshop on Scalable Computing for Big Data Analytics (SC-BDA 2012)*, 862–867. Piscataway, NJ: IEEE Computer Society.
9. Bryan, L. 2010. The social and psychological issues of high-density city space. In *Designing High-Density Cities for Social and Environmental Sustainability*, ed. E. Ng, 285–292. London: Earthscan.
10. Singapore Department of Statistics. 2013. Singapore in figures 2013. http://www.singstat.gov.sg/Publications/publications_and_papers/reference/sif2013.pdf.

11. Singapore Land Transport and Authority. 2013. Singapore land transport in brief 2013. http://www.lta.gov.sg/content/dam/ltaweb/corp/PublicationsResearch/files/FactsandFigures/Stats_in_Brief_2013.pdf.
12. Rahman, M., Li, X., and Veeravalli, B. 2012. Hybrid heuristic for scheduling data analytics workflow applications in hybrid cloud environments. In *Proceedings of the 2011 IEEE Symposium on Parallel and Distributed Processing Workshops and PhD Forum (IPDPSW'11)*, 966–974. Piscataway, NJ: IEEE Computer Society.
13. Hintjens, P. 2013. *ZeroMQ: Messaging for Many Applications*. Sebastopol, CA: O'Reilly.
14. Kim, H., el-Khamra, Y., Rodero, I., et al. 2011. Autonomic management of application workflows on hybrid computing infrastructure. *Scientific Computing* 19:75–89.
15. Gogouvitis, S., Konstanteli, K., Waldschmidt, S., et al. 2012. Workflow management for soft real-time interactive applications in virtualized environments. *Future Generation Computer Systems*, 28:193–209.
16. Fernandez, H., Tedeschi, C., and Priol, T. 2011. A chemistry-inspired workflow management system for scientific applications in clouds. In *Proceedings of the Seventh International Conference on e-Science (e-Science'11)*, 39–46. Piscataway, NJ: IEEE Computer Society.
17. Kim, H., el-Khamra, Y., Rodero, I., et al. 2011. Autonomic management of application workflows on hybrid computing infrastructure. *Scientific Programming* 19:75–89.

5

Migrating e-Science Applications to the Cloud: Methodology and Evaluation

Steve Strauch, Vasilios Andrikopoulos, Dimka Karastoyanova, and Karolina Vukojevic-Haupt

CONTENTS

Summary

Migrating an existing application to the cloud is a complex and multi-dimensional problem requiring in many cases adapting the application in significant ways. Taking a look in particular into the database layer of the application, this involves dealing with differences in the granularity of interactions, refactoring of the application to cope with remote data sources, and addressing data confidentiality concerns. In this chapter we introduce an application migration methodology that incorporates these aspects, and a decision support, application refactoring and data migration tool which supports application developers in realizing this methodology. We evaluate the proposed methodology and enabling tool using a case study conducted in the context of an e-science project.

5.1 Introduction

e-Science is an active field of research striving to enable faster scientific discovery and groundbreaking research in different scientific domains by means of information technology (IT). It is considered a new paradigm for science and is referred to as the *fourth paradigm* (Hey et al., 2009) or *data-intensive science*; it unifies theory, experiments, and simulation for data exploration for the purpose of scientific discovery. Existing literature shows that myriad available software systems, like Kepler, Triana, Taverna, Pegasus, and so on, support only some of the experiment life cycle phases and are applicable only for specific scientific domains (Taylor et al., 2006).

Due to its interdisciplinary nature, e-science exhibits a high degree of complexity, mainly due to the technical challenges and interoperability deficiencies of the existing software, the large amounts of data produced and consumed by the computational tools and systems, and the computational intensity and distributed characteristics of the IT environment observed in scientific computing. One major issue in current research is the integration of existing software and tools, across domains and organizational structures, for enabling the collaborative modeling of more complex scientific experiments and their execution. The most prominent approach for integrating software systems for the purpose of performing scientific experiments is workflow technology. Workflows are defined in terms of control flow among tasks comprising an experiment and the data exchanged among them (i.e., data flow). Moreover, the tasks in a workflow stand for a concrete unit of work that can be implemented by a computational, configuration, or visualization tool or by human users.

The available scientific workflow systems can be classified in two groups based on the fundamental features of the workflows they realize. There are data-driven scientific workflow systems, such as Kepler, Triana, Taverna, and Pegasus (Taylor et al., 2006), which stem from research in scientific computing. In such workflows, the focus is on modeling experiments in terms of how scientific data are processed (i.e., the tasks in a workflow are data-processing tasks), distributed, and placed on computing nodes in terms of computing jobs. There are also control flow-based scientific workflow systems, such as SimTech Scientific Workflow Management System (SWfMS; http://www.iaas.uni-stuttgart.de/forschung/projects/simtech/projects.php) and Trident (http://research.microsoft.com/en-us/collaboration/tools/trident.aspx), which support workflows with emphasis on the control flow among computational tasks, while the data consumed and produced by the software systems follow the control flow. In these workflows, the computational tasks are implemented by individual software systems, which in turn may distribute the computation over multiple computing nodes; however, this is kept transparent for the workflow system. The enacting environment, also called the workflow management system or workflow engine, is mainly dealing with orchestrating the software systems as well as human users. Such workflows have been developed as extensions to the available workflow technology from business applications.

These two different types of scientific workflow systems exhibit very different qualities of service characteristics, such as scalability, robustness, interoperability, reusability, and flexibility (De Roure, Goble, and Stevens, 2009; Görlach et al., 2011; Sonntag and Karastoyanova, 2010). The systems based on the conventional workflow technology from the business domain exhibit better quality-of-service characteristics. This can be explained mainly by the differences in the workflow metamodels and by the longer development, improvements, and evolution of workflow systems that took place in the field of enterprise application management (or the level of maturity reached by the workflow technology in this domain).

In recent years, cloud computing has gained significant acceptance in both the enterprise application management and scientific computing for its promise to reduce infrastructure costs and provide virtually unlimited computational power and data storage (Armbrust et al., 2009)—requirements of particular importance for businesses and of even greater importance to scientists and research organizations. While research in this field is active in providing novel concepts, techniques, and principles toward building cloud-native applications, there is a significant effort to cloud enable existing applications to reuse existing systems and therefore investments. Typically, cloud-enabling applications are related to the migration of whole systems or parts of them on a public or private cloud environment (Andrikopoulos et al., 2013; Deelman et al., 2008). Current research in migration methodologies and techniques, both specific to the e-science domain and outside it, is presented in Section 5.3.

In this work, we present a vendor- and technology-independent methodology for migrating the database layer of applications and refactoring the application architecture as positioned in existing methodologies for migration of applications (see Section 5.4). The methodology is applicable to applications in different application domains and is agnostic to the types of data sources. It fulfills requirements also presented in this work, which we have identified in collaboration with software engineers and domain experts in several research projects. We use this methodology to migrate the database layer of a scientific workflow management system (SimTech SWfMS), which we developed in the scope of our research activities in the SimTech project. The architecture and implementation details of the system, as well as the motivation for the database layer migration, are first presented in Section 5.2. The migration of the SimTech SWfMS has been done using the Cloud Data Migration Support Tool—a proof-of-concept implementation of the methodology. Both the introduced methodology and the supporting tool have been evaluated, and our findings are presented in Section 5.5. Our concluding remarks and plans for future work are presented in Section 5.6.

5.2 Motivating Scenario

As a motivating scenario from the e-science field, we use the integrated and interactive SWfMS developed in the context of the SimTech project (Sonntag, Hahn, and Karastoyanova, 2012; Sonntag and Karastoyanova, 2010). The SimTech SWfMS is a distributed system based on conventional workflow technology adapted to the needs of scientific workflows. The main components of the SimTech SWfMS are a modeling and monitoring tool, a workflow engine, an enterprise service bus, an auditing system, a messaging system, several database management systems, and an application server running the simulation services.

We present the architecture of the SimTech SWfMS in Figure 5.1. The user interacts with the system using the modeling and monitoring tool. SimTech SWfMS provides a graphical user interface to model, execute, and monitor scientific workflows. When the user initiates the execution of a workflow, the tool automatically deploys the workflow model on the workflow engine, which makes the simulation workflow available for use. The workflow can be instantiated as many times as needed. The instantiation of a scientific workflow is the beginning of the execution phase of the workflow life cycle.

The workflows executed by the workflow engine describe the ordered execution of different tasks such as data preparation, computation, or visualization. In our case, these tasks are realized by web services hosted on an application server. During the execution of a workflow, the workflow engine navigates along the predefined control flow and also interacts with these

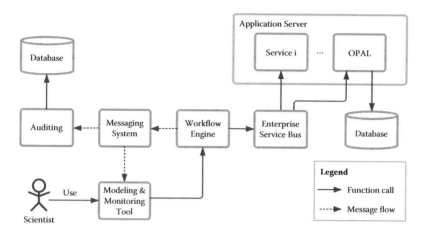

FIGURE 5.1
Main components of the SimTech SWfMS architecture.

web services through the service bus; that is, it sends a request for invocation of a web service and receives the results back from the web services. The service bus is also responsible for service discovery and selection if information about concrete services to be used is not available during the workflow deployment step.

The workflow engine also produces fine granular workflow execution events and publishes them to the messaging system. These events are consumed by the modeling and monitoring tool as well as by the auditing application. The modeling and monitoring tool uses the execution information to enable the live monitoring of running workflows. The auditing application captures the same execution information and saves it into a database to enable provenance and later analysis.

The actual workflow that serves as an example in the following is a *kinetic Monte Carlo simulation* (KMS) that invokes several web services as part of the simulation of solid bodies. These web services are implemented by modules of the OPAL application (Sonntag et al., 2011). During their operation, the OPAL web services access a MySQL database for both read and write operations. The example simulation of solid bodies is long running and requires significant computing power. Speeding up the simulation was a challenge that led to the decision to make use of cloud infrastructures, with the goal to acquire additional computational resources and data storage for the time of executing the simulation. This was indeed the major motivation for migrating the simulation workflow system or parts of it to the cloud.

Since the example simulation produces big amounts of data, one of our decisions was to temporarily migrate the database layer of the OPAL web services into the cloud, thus realizing the migration scenario *cloud bursting* (Strauch et al., 2013b), with Amazon Relational Database Service (RDS) as the migration target. However, migrating only the database layer to Amazon would result in extensive data transfer between the OPAL services on the premises and the database off the premises, therefore creating a potential bottleneck. Considering this, we decided also to migrate the business logic of the application to the cloud. Consequently, the modeling and monitoring tool was kept on the premises, while the remaining parts of the SimTech SWfMS were moved to an off-premises infrastructure. As a result, we not only avoid bottlenecks but also reduce costs since for most cloud providers data transfer inside the cloud is significantly cheaper than data transfer from and to the cloud.

The challenges we faced during this process were the following:

- which part of the system to migrate,
- what is the target system to migrate on,
- if and how to adapt the existing system to operate correctly after the migration,
- and most important, the lack of automated support with respect to these decisions.

To address these challenges, in this work we present a methodology that incorporates decision and refactoring support for migration of the database layer of applications to the cloud. For this purpose, in the following section we focus on investigating available methodologies and decision support systems (DSSs) for such scenarios.

5.3 Related Work

The state of the art we investigate in this section covers three aspects. First, we review existing literature on recommendations, benefits, and use cases with respect to the usage of cloud computing for e-science. Second, we investigate available vendor-specific and vendor-independent methodologies and guidelines for migrating either the database layer or the whole application to the cloud. Then, we consider available recommendations and DSSs with respect to migration to the cloud.

Mudge et al. reported an increase in speed by a factor of five on execution times when they migrated an e-science application from the domain of geophysics from on premises to the cloud, considering services from Amazon AWS and Microsoft Windows Azure (Mudge et al., 2011). Cala et al. used cloud computing to satisfy the demand for increased computation power and need for storing large volumes of data by migrating an existing e-science application for predicting chemical activity to Microsoft Windows Azure (Cala et al., 2013). The migration scenarios we are using in our methodology not only cover enterprise use cases but also cover scientific scenarios, as we have collaborated with industry partners and domain experts from the e-science domain while identifying them. Zinn et al. migrated an existing application based on scientific workflows from the domain of astronomy to Microsoft Windows Azure (Zinn et al., 2010). The existing application we migrate to the cloud for the purpose of evaluating our approach is also based on scientific workflows. Deelman et al. evaluated the cost of running e-science applications in the cloud, focusing on the trade-off between different workflow execution modes and provisioning plans, and came to the conclusion that the costs highly depend on the selected deployment strategy (Deelman et al., 2008). We do not explicitly consider costs but provide recommendations and guidelines with respect to the deployment strategy.

Amazon proposes a phase-driven approach consisting of six phases for migration of an application to its cloud infrastructure (Varia, 2010). The data migration phase is subdivided into a selection of the concrete Amazon AWS service and the actual migration of the data. Amazon provided recommendations regarding which of their data and storage services best fit for storing a specific type of data; for example Amazon Simple Storage Service (Amazon S3, http://aws.amazon.com/s3/) is ideal for storing large write-once, read-many types of objects. As the methodology proposed by Amazon focuses on Amazon AWS data and storage services only, we abstracted from this methodology and integrated the guidelines in our proposal. In addition to several product-specific guidelines and recommendations (Microsoft, 2013a, 2013b), Microsoft provided a Windows Azure SQL Database Migration Wizard (http://sqlazuremw.codeplex.com) and the synchronization service Windows Azure SQL Data Sync (http://www.windowsazure.com/en-us/manage/services/sql-databases/getting-started-w-sql-data-sync/). We reused some of these tools, tutorials, and wizards and refer to them during the data migration phase.

For the App Engine, Google is offering the tool Bulk Loader (http://bulkloadersample.appspot.com), which supports both the import of CSV and XML files into the App Engine Data Store and the export as CSV, XML, or text files. The potentially required transformations of the data during the import are customizable in configuration files. In addition, Google supports the user when choosing the appropriate data store or service and during its configuration (Google, 2013b). Moreover, they provide guidelines to migrate the whole application to Google App Engine (Google, 2013a). We refer to the tools during the migration phase and abstract from the vendor-specific guidelines and recommendations to integrate them in our tool.

Salesforce provides data import support to their infrastructure via a web user interface or the desktop application Apex Data Loader (http://sforce-app-dl.sourceforge.net). Another option to migrate and integrate with cloud providers such as Salesforce is to hire external companies that specialize in migration and integration, such as Informatica Cloud (http://www. informaticacloud.com). In addition to the tools or external support, Salesforce provides data migration guidelines (salesforce.com, 2013). We consider the non-Salesforce-specific steps for our proposed methodology. As discussed extensively in Section 5.4, Laszewski and Nauduri also proposed a vendor-specific methodology for the migration to Oracle products and services by providing a detailed methodology, guidelines, and recommendations focusing on relational databases (Laszewski and Nauduri, 2011). We base our proposal on their methodology, by abstracting from it and adapting and extending it.

Apart from the vendor-specific migration methodologies and guidelines, there are also proposals independent from a specific cloud provider. Reddy and Kumar proposed a methodology for data migration that consists of the following phases: design, extraction, cleansing, import, and verification. Moreover, they categorized data migration into storage migration, database migration, application migration, business process migration, and digital data retention (Reddy and Kumar, 2011). In our proposal, we focus on the storage and database migration as we address the database layer. Morris specifies four golden rules of data migration with the conclusion that the IT staff does not often know about the semantics of the data to be migrated, which causes a lot of overhead effort (Morris, 2012). With our proposal of a step-by-step methodology, we provide detailed guidance and recommendations on both data migration and required application refactoring to minimize this overhead. Tran et al. adapted the function point method to estimate the costs of cloud migration projects and classified the applications potentially migrated to the cloud (Tran et al., 2011). As our assumption is that the decision to migrate to the cloud has already been taken, we do not consider aspects such as costs. We abstract from the classification of applications to define the cloud data migration scenarios and reuse distinctions, such as complete or partial migration to refine a chosen migration scenario.

As we discuss the prototypical realization of a tool providing support and guidelines while deciding for a concrete cloud data store or service, the migration, and the refactoring of the application architecture accordingly, in the following we also investigate the state of the art on decision support systems (DSSs) (Power, 2002) in the area of cloud computing. Khajeh-Hosseini et al. introduced two tools that support the user when migrating an application to infrastructure-as-a-service (IaaS) cloud services (Khajeh-Hosseini et al., 2011). The first one enables the cost estimation based on a UML deployment model of the application in the cloud. The second tool helps to identify advantages and potential risks with respect to the cloud migration. None of these tools is publicly available. We do not consider the estimation of costs or the identification of risks as our assumption is that the decision for migration to the cloud has already been taken. We consider aspects such as costs, business resiliency, effort, and so on to be considered before following our methodology and using the tool (Andrikopoulos et al., 2013). Menzel and Ranjan developed CloudGenius, a DSS for the selection of an IaaS cloud provider focusing on the migration of web servers to the cloud based on virtualization technology (Menzel and Ranjan, 2012). As we provide support for the migration of the database layer, we focus on another type of middleware technology. Our approach is also not limited to a specific cloud service delivery model and migration by using virtualization technology.

5.4 Migration Methodology and Tool Support

As discussed, in this section we introduce a step-by-step methodology for the migration of the database layer to the cloud and the refactoring of the application architecture. Before we introduce the methodology, we investigate the requirements to be fulfilled by such a methodology.

5.4.1 Requirements

The *functional* and *nonfunctional* requirements we present in this section aim to provide decision support and guidelines for both migrating an application database layer to the cloud and refactoring of the application architecture. The presented requirements have been identified during our work on various research projects, especially during our collaboration with industry partners and IT specialists from the e-science domain.

5.4.1.1 Functional Requirements

The following functional requirements (FRs) must be fulfilled by any methodology for migration of the database layer to the cloud and refactoring of the application architecture:

FR_1 *Support of Data Stores and Data Services*: The methodology must support the data migration for both fine- and coarse-grained types of interactions (e.g., through SQL and service APIs, respectively).

FR_2 *On-Premises and Off-Premises Support*: The methodology has to support data stores and data services that are either hosted on the premises or off the premises and using both cloud and noncloud technologies.

FR_3 *Independence from Database Technology*: The methodology has to support both established relational database management systems (Codd, 1970) and NoSQL data stores (Sadalage and Fowler, 2012) that have emerged in recent years.

FR_4 *Management and Configuration*: Any tool supporting such a methodology must provide management and configuration capabilities for data stores, data services, and migration projects bundling together different migration actions. This includes, for example, the registration of a new data store, including its configuration data (e.g., database schemas, database system end point uniform resource locators [URLs], etc.). It must also support the creation of new migration projects for documentation of the decisions and actions taken during migration.

FR$_5$ *Support for Incompatibility Identification and Resolution*: Any potential incompatibilities
        (e.g., between SQL versions supported by different data services) must be identified,
        and guidance must be provided on how to overcome them. For this purpose, the
        methodology has to incorporate the specification of functional and nonfunctional
        requirements for both the (source) database layer used before the migration and the
        target data store or data service.

FR$_6$ *Support for Various Migration Scenarios*: As the data migration depends on the context
        and the concrete use case (e.g., backup, archiving, or cloud bursting), the methodology
        has to support various migration scenarios.

FR$_7$ *Support for Refactoring of the Application Architecture*: The amount of refactoring of the
        application architecture during the migration of the database layer to the cloud depends
        on many aspects, such as the supported functionalities of the target data store or data
        service, use case, and so on. It is therefore required that the methodology provides
        guidance and recommendations on how to refactor the application architecture.

5.4.1.2 Nonfunctional Requirements

In addition to the required functionalities, a methodology for migration of the database layer to the cloud and refactoring of the application
architecture should also respect the following properties of nonfunctional
requirements (NFRs):

NFR$_1$ *Security*: Both data export from a source data store and data import to a target data
        store require confidential information such as data store location and access
        credentials. Any tool supporting the methodology should therefore consider
        necessary authorization, authentication, integrity, and confidentiality mechanisms
        and enforce user-wide security policies when required.

NFR$_2$ *Reusability*: As the migration of data can be seen as either the migration of only the
        database layer or as part of the migration of the whole application, the methodology
        has to be reusable with respect to the integration into a methodology for migration
        of the whole application to the cloud, such as the one proposed by Varia for Amazon
        (Varia, 2010).

NFR$_3$ *Extensibility*: The methodology should be extensible to incorporate further aspects that
        have an impact on the data migration to the cloud, such as regulatory compliance.
        For example, in the United States, the cloud service provider is responsible for
        ensuring compliance to regulations (Louridas, 2010), but in the European Union, it is
        the cloud customer that is ultimately responsible for investigating whether the
        provider realizes the Data Protection Directive (Cate, 1994).

5.4.2 Migration Methodology

The step-by-step methodology we introduce in this section refines and
adapts the migration methodology proposed by Laszewski and Nauduri
(2011) to address the identified requirements. The methodology (Laszewski
and Nauduri, 2011) consists of seven distinct phases (Figure 5.2). During
the *Assessment* phase, information relevant for project management, such as
drivers for migration, migration tools, and migration options, is collected

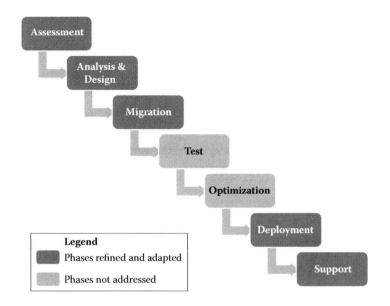

FIGURE 5.2
Migration methodology as proposed by Laszewski and Nauduri in 2011, with supported phases highlighted. (Redrawn from Laszewski, T., and P. Nauduri (2011). *Migrating to the Cloud: Oracle Client/Server Modernization*. New York: Elsevier.)

to assess the impact of the database migration on the IT ecosystem. The *Analysis-and-Design* phase investigates the implementation details on the target database (e.g., potentially different data types and transaction management mechanisms being used). The goal of this phase is the creation of a plan to overcome potential incompatibilities between the source and target data store while avoiding changes in the business logic of the application. The *Migration* phase deals with the migration of the data from the source data store to the target data store in a testing environment, including tasks such as database schema migration, database stored procedures migration, and data migration. After the migration, both the database and the application have to be tested in the *Test* phase. This includes, for example, tasks such as data verification and testing the interaction of the application with the new target data store. As applications are in general highly optimized for a particular database, after the migration to another target data store the performance might be poor. Thus, optimizations based on the new target store used are applied in the *Optimization* phase to improve the performance. The goal of the *Deployment* phase is to deploy the final system, including actually migrating the database, to the production environment.

At first glance, the methodology of Laszewski and Nauduri addresses most of the requirements discussed previously. However, it discusses its phases on a high level that is not suitable for direct application, requiring further refinement in practice. Furthermore, it fails to satisfy some of the most important requirements that we identified. More specifically, as the methodology focuses on Oracle solutions, it only considers the relational database management system of Oracle as the target data store and the following relational data stores as the source databases for the migration: Microsoft SQL Server (http://www.microsoft.com/en-us/sqlserver), Sybase (http://www.sybase.com), IBM DB2 (http://www.ibm.com/software/data/db2), and IBM Informix (http://www.ibm.com/software/data/informix/). All of these databases are data stores supporting fine-grained interactions through SQL. It is unclear whether the methodology also supports data services because no information can be found on this aspect in Laszewski and Nauduri's work (2011) (FR_1). The methodology is not independent from the database technology as it focuses on a small set of relational databases and does not support NoSQL approaches (FR_3). Moreover, the methodology is limited to the pure outsourcing of the database layer to the cloud and does not consider the context and specifics of migration scenarios such as cloud bursting, backup, and archiving (FR_6). As concrete migration scenarios are not considered, their specifics and the context cannot be considered for the guidance and recommendation toward refactoring of the application architecture. In addition, the guidance and recommendations for the required adaptations of the application architecture during the migration are limited since the migration methodology (Laszewski and Nauduri, 2011) considers only one vendor-specific relational target data store and a small subset of vendor-specific relational data stores as the source data store (FR_7). The vendor specificity also has the consequence that the methodology does not consider the reusability aspect with respect to the integration or combination of this methodology with other existing proposals for migration to the cloud (NFR_2).

To address these deficiencies, in the following we propose a vendor- and database technology-independent step-by-step methodology that refines and adapts the one proposed by Laszewski and Nauduri (2011). Figure 5.2 provides an overview of the phases of the methodology proposed that we adapted and refined. Figure 5.3 provides an overview of our proposal consisting of seven steps. All steps are semiautomatic, in the sense that a human (e.g., the application developer in charge of the migration) has to provide input and follow the recommendations and guidelines provided by the methodology. Figure 5.3 also shows the mapping between the proposed methodology and the one in Laszewski and Nauduri's 2011 work. As can be seen, no direct support for the Test and Optimization phases is provided by our proposal since there are no identified requirements explicitly requiring these phases. The impact of not supporting these phases is evaluated in Section 5.5. The steps of the methodology are discussed next.

5.4.2.1 Step 1: Select Migration Scenario

The first step in our proposed methodology is the *selection of the migration scenario*. For this purpose, we use the 10 *Cloud Data Migration Scenarios*

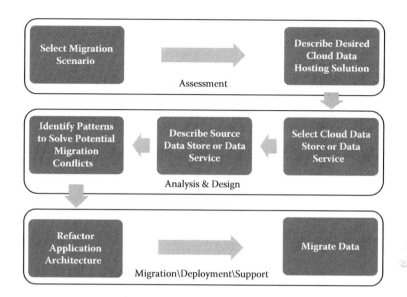

FIGURE 5.3
Methodology for migration of the database layer to the cloud and refactoring of the application architecture.

identified by Strauch et al. (2013b): database layer outsourcing; use of highly scalable data stores; geographical replication; sharding; cloud bursting; working on data copy; data synchronization; backup; archiving; and data import from the cloud (FR_6). These migration scenarios cover both migration directions between on the premises and off the premises (FR_2).

Based on the selection of the migration scenario, a *migration strategy* is formulated by considering properties such as live or nonlive migration, complete or partial migration, and permanent or temporary migration to the cloud. During this step, potential conflicts between the migration scenario selected and the refined migration strategy should be explicitly addressed by proposing solutions to the user (e.g., the choice of a different migration scenario). An example of a conflict is the selection of the migration scenario cloud bursting and the choice of a permanent migration to the cloud in the strategy. The purpose of this migration scenario is by definition to migrate the database layer to the cloud to cover peak loads and migrate it back afterward; choosing permanent migration as part of the strategy therefore cannot be satisfied.

5.4.2.2 Step 2: Describe Desired Cloud Data Hosting Solution

The specification of functional and nonfunctional requirements with respect to the target data store or data service is the focus of the second step. We define the *cloud data hosting solution* as the concrete configuration of a cloud data store or cloud data service in terms of a set of concrete functional and nonfunctional properties (FR_1). Therefore, we derived an initial set of

TABLE 5.1

Excerpt of Categories and Properties for Specification of Requirements of Cloud Data
Hosting Solutions

Categories	Properties	Available Options
Scalability	Degree of automation	Manual, automated
	Type	Horizontal, vertical
	Degree	Virtually unlimited, limited
	Time to launch new instance	None, duration in minutes
Availability	Replication	Yes, no
	Replication type	Master-slave, master-master
	Replication method	Synchronous, asynchronous
	Replication location	Same data center
		Different data center (same region)
	Automatic failover	Yes, no
	Degree	99.9%, 99.999%
Security	Storage encryption	Yes, no
	Transfer encryption	Yes, no
	Firewall	Yes, no
	Authentication	Yes, no
	Confidentiality	Yes, no
	Integrity	Yes, no
	Authorization	Yes, no
Interoperability	Data portability	None, import, export
		One-way synchronization
	Data exchange	XML, JSON, proprietary
	Format	
	Storage access	SOA, REST-API, SQL, proprietary
	ORM	JPA, JDO, LINQ
	Migration and deployment support	Yes, no
	Supported IDE	Eclipse, NetBeans, IntelliJ IDEA
	Developer SDKs	Java, .Net, PHP, Ruby
Storage	Storage type	RDBMS, NoSQL
CAP	Consistency model	Strong, weak, eventual
	Availability in case of partitioning	Available, not available

properties grouped into different categories based on the analysis of current
data store and data service offerings of established cloud providers such as
Amazon, Google, and Microsoft. Table 5.1 provides an excerpt of the categories and corresponding properties we considered. These categories cover
both relational and NoSQL solutions (FR_3, FR_5).

5.4.2.3 Step 3: Select Cloud Data Store or Data Service

The *concrete target data store or data service* for the migration is selected in
step 3 by mapping the properties of the cloud data hosting solution specified

in the previous step to the set of available data stores and data services that have been categorized according to the same nonfunctional and functional properties. Implementing this step requires data stores and data services to be previously specified according to the set of functional and nonfunctional properties either directly by the cloud providers or by the users of the methodology. The management and configuration capabilities required for this specification, however, can be used at a later time to also make new cloud data stores and data services available (FR$_4$).

5.4.2.4 *Step 4: Describe Source Data Store or Data Service*

As it is not sufficient to consider only where the data has to be migrated, in step 4 the *functional and nonfunctional properties of the source data store or data service* are also described to identify and solve potential migration conflicts, such as the database technology used or whether the location is on or off the premises (FR$_5$).

5.4.2.5 *Step 5: Identify Patterns to Solve Potential Migration Conflicts*

The use of cloud technology leads to challenges such as incompatibilities with the database layer previously used or the accidental disclosing of critical data (e.g., by moving them to the public cloud). Incompatibilities in the database layer may refer to inconsistencies between the functionalities of an existing traditional database layer and the characteristics of an equivalent cloud data hosting solution. Therefore, in the fifth step conflicts are identified by checking the compatibility of the properties of the target data store selected in step 3 with the properties of the source data store or service used before the migration (FR$_5$). As a way to address these conflicts, in previous work (Strauch et al., 2013c) we have defined a set of *cloud data patterns* as the best practices to deal with them that can be reused here.

5.4.2.6 *Step 6: Refactor Application Architecture*

As the migration of the database layer also has an impact on the remaining application layers (presentation and business logic; Fowler et al., 2002), the methodology should provide guidelines and hints on what should be considered for the refactoring of the application. Special focus should be given to the adaptation of the network, the data access layer, and the business logic layer of the application, depending on the outcomes of the previous steps (FR$_7$). Networking adaptation might require, for example, the reconfiguration of open ports in the enterprise firewall. Although the cloud data store might be fully compatible with the data store previously used, the migration requires at least a change to the database connection string in the data access layer. The impact of the database layer migration to the cloud on the business logic layer depends on several aspects, such as the migration scenario and the incompatibilities of the source and target data store. In case of switching from a relational database to a NoSQL data service, the business logic needs to be significantly adapted as the characteristics of these two technologies are different, for example, with respect to transaction support, relational database schema versus schema-free or schema-less NoSQL solution, and quality of services (Sadalage and Fowler, 2012).

5.4.2.7 Step 7: Migrate Data

The final step, *migrating the data*, entails the configuration of the connections to the source and target data stores or services by requiring input on the location, credentials, and so on from the user. This step should also provide *adapters* for the corresponding source and target stores, bridging possible incompatibilities between them, or reuse of the data export and import tools offered by the different cloud providers. As the last step is dealing with potentially confidential information, to prevent other users from accessing the data, a tool supporting the proposed methodology has to support the required security mechanisms (NFR_1).

5.4.3 Realization

In this section, we introduce the realization of a *Cloud Data Migration Tool* for the migration of the database layer to the cloud and the refactoring of the application architecture (Strauch et al., 2013a). More specifically, to support the proposed methodology, the Cloud Data Migration Tool provides two main functionalities. On the one hand, it provides a repository for cloud data stores and cloud data services and allows browsing through it, even without user registration. In addition, it implements the required management functionality to add new entries in the repository by specifying their functional and nonfunctional properties. On the other hand, the tool guides the user through the first six steps of the proposed methodology through a DSS. For the last step of migrating the data, the tool is equipped with adapters that allow the automatic export of data from the source data store and their import in the target data store. Currently, the tool has source adapters for PostgreSQL (http://www.postgresql.org) and Oracle MySQL (http://www.mysql.com). We provide target adapters for a number of cloud data stores and data services, such as Amazon RDS (http://aws.amazon.com/rds/) and 10gen MongoDB (http://www.mongodb.org), MySQL in Amazon EC2 instances (http://aws.amazon.com/ec2/), Google Cloud SQL (http://cloud.google.com/products/cloud-sql/), and Amazon SimpleDB (http://aws.amazon.com/simpledb/). In addition to the adapters, the user is referred to various guidelines and tutorials provided by the different cloud providers (e.g., Google, 2013c). This is especially useful if no appropriate adapter is available for a particular data store or service.

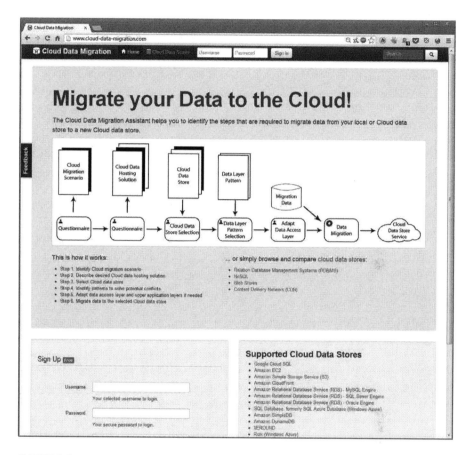

FIGURE 5.4
Screenshot of the realization of the cloud data migration tool.

Figure 5.4 provides an overview of the main page of the Cloud Data Migration Tool publicly available for free use (http://www.cloud-data-migration.com). As the user has to provide confidential data following the guidelines and recommendations of the tool (e.g., access credentials to the source and target data stores or services for data export and import in the last step), the user has to register with user, password, and e-mail address. After a migration project is finalized, the user can print a report of the decisions made during the migration, the identified conflicts, and their resolutions for the purpose of documentation and support. Currently, we are supporting the migration from one source data store to one target data store or service, and one migration project has to be created per migration. Extending the tool in order to support more than one target data stores per migration project is ongoing work.

The Cloud Data Migration Tool is realized as a Java 6 web application and follows a three-layer architecture. The presentation layer is realized using

HTML (hypertext markup language), JavaScript, JSP, and CSS. The business logic layer is implemented in Java. For the object-relational mapping, we use Java Data Objects version 3.1 and its implementation DataNucleus version 3.0 (http://www.datanucleus.org). For online hosting of the tool we use Google Cloud SQL as the data layer and run the whole application in Google's App Engine. A stand-alone, offline version of the tool also exists, allowing the user to run the tool locally. In this case, MySQL 5.5 is used for the data layer and Apache Tomcat version 7 as the servlet container. Further information is available on the website of the Cloud Data Migration Tool (http://www.cloud-data-migration.com).

5.5 Evaluation

In this section, we evaluate both the methodology introduced in Section 5.4.2 and the Cloud Data Migration Tool supporting this methodology presented in the previous section. For this purpose, we used the motivating scenario discussed in Section 5.2 as a case study involving the migration of the database layer of the SimTech SWfMS to the cloud.

As our investigation of the literature did not result in a method that specifically aims at the evaluation of migration methodologies, we focused our analysis on related evaluation methods and standards for software processes and software quality. Al-Qutaish and Berander et al. provided an overview of available software quality models and standards (Al-Qutaish, 2010; Berander et al., 2005). Based on their findings, we selected the International Organization for Standardization/International Electrotechnical Commission (ISO/IEC) 9126 standard provided for the evaluation of the Cloud Data Migration Tool, as its quality attribute model includes the metrics we considered most relevant, such as understandability and operability (Jung, Kim, and Chung, 2004). For the evaluation of software processes, there are multiple guidelines (e.g., Shull, Carver, and Travassos, 2001; Sommerville, 1996) and standardized best practices, such as Capability Maturity Model Integration (CMMI) (CMMI Product Team, 2010) and the Continual Service Improvement (CSI) module of the IT Infrastructure Library (ITIL) (Case and Spalding, 2011). We based our evaluation of the migration methodology on the ITIL CSI process but adapted it to consider the technical aspects of the methodology by considering appropriate metrics for software processes provided by Daniel (2004). A simplified representation of the resulting process is shown in Figure 5.5.

In the first step, a strategy for the realization of the process was determined. In this case, our strategy was to use the Cloud Data Migration Tool discussed in the previous section in conjunction with a specific migration

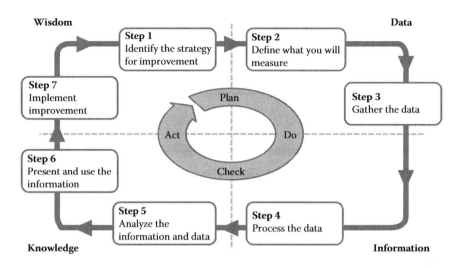

FIGURE 5.5
CSI seven-step process used for the evaluation. (Adapted from Case, G., and G. Spalding (2011). *ITIL Continual Service Improvement.* London: TSO, The Stationery Office.)

scenario and investigate whether it supported the scenario in an effective and efficient manner. In the second step, which data will be collected needed to be defined. These data were the basis for the subsequent process steps. In our evaluation, we collected both qualitative and quantitative data. With respect to the former, we recorded the user-identified problems that occurred during the execution of the SimTech SWfMS migration as the means to evaluate the software quality of the Cloud Data Migration Tool. Such problems are gathered only in a qualitative manner (i.e., we were not interested in the number of problems that occurred but in a comprehensive description and classification of these problems). This approach increased the effort to gather the data but in turn enabled a more detailed and potentially more meaningful analysis. In terms of quantitative data, we recorded the time required for executing the various migration phases. To be able to compare our proposal with the one by Laszewski and Nauduri (2011), we chose to use their phases as the metric of the efficiency of our proposed approach. In this manner, we could attribute time elapsed to higher-level activities in addition to evaluating the impact of not incorporating the testing and optimization phases in our proposal.

To enable structured gathering and recording of problems that occurred, we defined a set of attributes related to them. Table 5.2 shows an example of such a problem that was identified during our evaluation and the information we collected for it. Every problem has a unique identifier (*ID*) and a descriptive *Name*. The attribute *Class* is used to classify the problem in predefined categories. We derived these categories from ISO/IEC 9126-1, which defines

TABLE 5.2

Documentation of an Identified Problem

ID	B7
Name	Connection failed
Class	Tool (operability)
Severity	High
Description	Although correct users with the required administrative roles existed in the MySQL database in the cloud, the application could not connect to the database.
Error handling	We were going through all the security (user and privilege) settings in the MySQL Workbench.
Solution	We set *max queries, max updates, max connections* to a value greater than zero for each user.
Adaptation	The user should obtain information about the limitations for the different accounts (users).

a quality model for software by subdividing software quality in different characteristics and subcharacteristics (Jung, Kim, and Chung, 2004). In our evaluation, we focused on the characteristics *functionality* and *usability* of the examined tool, in particular on the subcharacteristics *suitability* (for the former) and *understandability* and *operability* (for the latter), which are the possible values for the Class attribute. The problem identified in Table 5.2, for example, is classified under the operability subcharacteristic of usability. The attribute *Severity* describes the severity of a problem with respect to the impact on the migration result. The allowed values are *low, middle, high,* or *critical*. A detailed description of a problem is given with the attribute *Description*. The attribute *Error Handling* describes how the user has proceeded to find a solution for the problem that occurred. *Solution* describes how the problem was fixed. To eliminate the cause of the problem, adaptations of the tool may be needed; these are described by the attribute *Adaptation*.

In the third step, the actual gathering of data was performed. Using the Cloud Data Migration Tool, we migrated the database layer used by the OPAL web services to the cloud. The selected use case can be mapped to the migration scenario *Cloud Bursting* (Strauch et al., 2013b), with Amazon RDS as the migration target. Throughout all phases of the migration, we recorded any occurring problems, as shown in Table 5.2. In addition, we measured the time spent per migration phase supported by our step-by-step methodology (i.e., Assessment; Analysis and Design; and Migration, Deployment, and Support), as well as the time spent on testing. No optimization activity was implemented as part of the case study. In the fourth step of the evaluation, the previously gathered data were processed to organize and structure for further analysis. As we had already gathered the data in a structured and uniform manner (as described in step 2), further processing was not necessary.

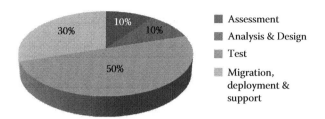

FIGURE 5.6
Amount of time spent per migration phase.

In the fifth step, the analysis of the gathered and processed data took place. Altogether, we recorded seven problems. Five of the recorded problems had a high priority; the remaining two had middle priority. Two of the occurred problems were due to bugs in the graphical user interface of the tool, one with middle and one with high priority. Two problems were caused by missing features, also one with middle and one with high priority. The rest of the problems, all with high priority, were caused by lack of appropriate information available to the user, as in the example of Table 5.2. The analysis of the identified problems with respect to their priority and the cause of the problems showed that the main weakness of the Cloud Data Migration Tool was a lack of information provided to the user. Further improvements toward this direction are therefore required in the future.

The analysis of the time spent per migration phase is summarized in Figure 5.6. As shown in the figure, half of the time was actually spent in the Test phase, which as explained in Section 5.4 is not directly supported by our methodology (and therefore also not by the Cloud Data Migration Tool). While this identifies a deficiency in our proposal, it can also be attributed at least in part to the acceleration of the other phases by the use of the Cloud Data Migration Tool. In any case, what can be identified is a clear need for the incorporation of the remaining two phases (Test and Optimization) in our methodology and as a result their support by the Cloud Data Migration Tool.

Finally, for the implementation of steps 6 and 7 of the ITIL CSI process (presentation and use of the information and implement improvements, respectively), we are currently in the process of incorporating the lessons learned by this case study in further research work.

5.6 Conclusions

The popularity of cloud computing has led to significant research in cloud-enabling applications, that is, migrating whole systems or only parts of them to the cloud. The e-science domain, especially the scientific workflow community, has reported concrete benefits from utilizing cloud infrastructures for isolated use cases. In this respect, there is a clear need for a methodology supporting the migration of e-science applications to the cloud. There are two key aspects that characterize e-science applications: large amounts of data and intensive computational tasks to be performed on these data. In this work, we focused on the former, discussing how to support the migration of the database layer of e-science applications (and beyond) to the cloud.

Supporting the migration of the database layer of an application to the cloud involves not only considering the requirements on the appropriate data source or service imposed by the application but also the possible need for adapting the application to cope with incompatibilities. In the previous sections, we presented a step-by-step methodology that considers both aspects of the migration. To construct this methodology, we first identified a series of functional and nonfunctional requirements from both e-science and business domains. We then adapted the methodology discussed by Laszewski and Nauduri (2011) to satisfy the identified requirements, resulting in our proposal for a seven-step end-to-end methodology for the migration of the database layer to the cloud and for the application refactoring required as part of this process.

Then, we discussed the realization of our proposal as a publicly available and free Cloud Data Migration Tool. The tool provides two fundamental functionalities: decision support in selecting an appropriate data store or service and refactoring support during the actual migration of the data. Users of the tool can currently create migration projects, define their requirements in terms of the migrated database layer to the cloud, describe their current database layer, and receive recommendations, hints, and guidelines on where and how to migrate their data. Conflict resolution is based on previously identified cloud data patterns, and data adapters are provided, allowing for the automatic migration of data to recommended data stores and services. We evaluated our proposal by migrating the SimTech SWfMS to Amazon Web Services solutions and showed that, while useful, our methodology and tool need further improvements.

In particular, according to our evaluation, our proposal needs to be extended to provide explicit support for the testing phase of the migration. The Cloud Data Migration Tool must be extended to provide sandboxing capabilities and both functional testing for bug fixing and performance benchmarking tools for different application workloads. These capabilities can also be used toward supporting the optimization of the database layer after its migration. Additional functionalities that are currently being implemented to the Cloud Data Migration Tool, as identified in the previous sections, include addressing the impact of the migration to compliance, supporting more than one source or target data stores or services and multiple migrations per project, increasing the number of adapters available in the tool, as well as improving the usability of the tool for scientists.

List of Abbreviations

API	Application Programming Interface
CSS	Cascading Style Sheets
IDE	Integrated Development Environment
JDO	Java Data Objects
JPA	Java Persistence API
JSON	JavaScript Object Notation
JSP	JavaServer Pages
LINQ	Language Integrated Query
NoSQL	Not Only Structured Query Language
ORM	Object-Relational Mapping
RDBMS	Relational Database Management System
REST	Representational State Transfer
SDK	Software Development Kit
SOA	Service-Oriented Architecture
SQL	Structured Query Language
UML	Unified Modeling Language
XML	Extensible Markup Language

Acknowledgments

The research leading to these results received funding from the European Union's Seventh Framework Program (FP7/2007–2013) projects 4CaaSt (http://www.4caast.eu; grant agreement no. 258862) and ALLOW Ensembles (http://www.allow-ensembles.eu; grant agreement no. 600792) and from the German Research Foundation (DFG) within the Cluster of Excellence in Simulation Technology (http://www.simtech.uni-stuttgart.de; EXC 310/1) at the University of Stuttgart.

Bibliography

Al-Qutaish, R. E. (2010). Quality models in software engineering literature: an analytical and comparative study. *Journal of American Science* 6(3), 166–175.

Andrikopoulos, V., T. Binz, F. Leymann, and S. Strauch (2013). How to adapt applications for the cloud environment. *Computing* 95(6), 493–535.

Armbrust, M., A. Fox, R. Griffith, et al. (2009). *Above the Clouds: A Berkeley View of Cloud Computing*. Technical Report UCB/EECS-2009-28. Berkeley: EECS Department, University of California, Berkeley.

Berander, P., L.-O. Damm, J. Eriksson, et al. (2005). *Software Quality Attributes and Trade-Offs*. Technical report. Karlskrona, Sweden: Blekinge Institute of Technology.

Cala, J., H. Hiden, S. Woodman, and P. Watson (2013). Cloud computing for fast prediction of chemical activity. *Future Generation Computer Systems* 29(7), 1860–1869.

Case, G., and G. Spalding (2011). *ITIL Continual Service Improvement*. London: TSO, The Stationery Office.

Cate, F. (1994). The EU Data Protection Directive, information privacy, and the public interest. *Iowa Law Review* 80, 431.

CMMI Product Team (2010). CMMI for Development, Version 1.3 (CMU/SEI-2010-TR-033). Software Engineering Institute, Carnegie Mellon University. http://www.sei.cmu.edu/library/abstracts/reports/10tr033.cfm.

Codd, E. F. (1970). A relational model of data for large shared data banks. *Communications of the ACM* 13(6), 377–387.

Daniel, G. (2004). *Software Quality Assurance: From Theory to Implementation*. Upper Saddle River, NJ: Pearson Education.

Deelman, E., G. Singh, M. Livny, B. Berriman, and J. Good (2008). The cost of doing science on the cloud: the montage example. In *Proceedings of the 2008 ACM/IEEE Conference on Supercomputing*, pp. 50:1–50:12. New York: IEEE Press.

De Roure, D., C. Goble, and R. Stevens (2009). The design and realisation of the My Experiment Virtual Research Environment for Social Sharing of Workflows. *Future Generation Computer Systems* 25, 561–567.

Fowler, M., et al. (2002, November). *Patterns of Enterprise Application Architecture*. Boston: Addison-Wesley Professional.

Google (2013a). Google App Engine—migrating to the high replication datastore. http://developers.google.com/appengine/docs/adminconsole/migration.

Google (2013b). Google App Engine—uploading and downloading data. http://developers.google.com/appengine/docs/python/tools/uploadingdata?hl=en.

Google (2013c). Google Cloud SQL—importing and exporting data. http://developers.google.com/cloud-sql/docs/import_export.

Görlach, K., M. Sonntag, D. Karastoyanova, F. Leymann, and M. Reiter (2011). Conventional workflow technology for scientific simulation. In *Guide to e-Science*, pp. 323–352. New York: Springer.

Hey, A. J., S. Tansley, K. M. Tolle, et al. (2009). *The Fourth Paradigm: Data-Intensive Scientific Discovery*. Redmond, WA: Microsoft Research.

Jung, H.-W., S.-G. Kim, and C.-S. Chung (2004). Measuring software product quality: a survey of ISO/IEC 9126. *IEEE Software* 21(5), 88–92.

Khajeh-Hosseini, A., I. Sommerville, J. Bogaerts, and P. Teregowda (2011). Decision support tools for cloud migration in the enterprise. In *Proceedings of CLOUD'11*, pp. 541–548. New York: IEEE Press.

Laszewski, T., and P. Nauduri (2011). *Migrating to the Cloud: Oracle Client/Server Modernization*. New York: Elsevier.

Louridas, P. (2010). Up in the air: moving your applications to the cloud. *IEEE Software* 27(4), 6–11.

Menzel, M., and R. Ranjan (2012). CloudGenius: decision support for web server cloud migration. In *Proceedings of WWW'12*, pp. 979–988. New York: ACM.

Microsoft (2013a). Develop and deploy with Windows Azure SQL Database. http://social.technet.microsoft.com/wiki/contents/articles/994.develop-and-deploy-with-windows-azure-sql-database.aspx.

Microsoft (2013b). Guidelines and limitations (Windows Azure SQL Database). http://msdn.microsoft.com/en-us/library/windowsazure/ff394102.aspx.

Morris, J. (2012). *Practical Data Migration*, 2nd ed. London: BCS, The Chartered Institute for IT.

Mudge, J., P. Chandrasekhar, G. Heinson, and S. Thiel (2011). Evolving inversion methods in geophysics with cloud computing—a case study of an escience collaboration. In *Proceedings of e-Science'11*, pp. 119–125. Stockholm, Sweden: IEEE.

Power, D. (2002). *Decision Support Systems: Concepts and Resources for Managers*. Quorum Books.

Reddy, V. G., and G. S. Kumar (2011). Cloud computing with a data migration. *Journal of Current Computer Science and Technology* 1 (06).

Sadalage, P. J., and M. Fowler (2012). *NoSQL Distilled: A Brief Guide to the Emerging World of Polyglot Persistence*. Boston: Addison-Wesley.

salesforce.com (2013). Salesforce helpdata importing overview. http://help.salesforce.com/HTViewHelpDoc?id=importing.htm&language=en_US.

Shull, F., J. Carver, and G. H. Travassos (2001). An empirical methodology for introducing software processes. *SIGSOFT Software Engineering Notes* 26(5), 288–296.

Sommerville, I. (1996). Software process models. *ACM Computing Surveys* 28(1), 269–271.

Sonntag, M., M. Hahn, and D. Karastoyanova (2012, September). Mayflower—explorative modeling of scientific workflows with BPEL. In *Proceedings of CEUR Workshop'12*, pp. 1–5. New York: Springer.

Sonntag, M., S. Hotta, D. Karastoyanova, D. Molnar, and S. Schmauder (2011). Using services and service compositions to enable the distributed execution of legacy simulation applications. In *Towards a Service-Based Internet*, pp. 242–253. New York: Springer.

Sonntag, M., and D. Karastoyanova (2010). Next generation interactive scientific experimenting based on the workflow technology. In *Proceedings of MS'10*, Alhajj, R. S., Leung, V. C. M., Saif, M., and Thring, R., editors. pp. 349–356. Banff, Alberta, Canada.

Strauch, S., V. Andrikopoulos, T. Bachmann, D. Karastoynova, S. Passow, and K. Vukojevic-Haupt (2013a, December). Decision support for the migration of the application database layer to the cloud. In *Proceedings of CloudCom'13*. Washington, DC: IEEE Computer Society Press.

Strauch, S., V. Andrikopoulos, T. Bachmann, and F. Leymann (2013b). Migrating application data to the cloud using cloud data patterns. In *Proceedings of CLOSER'13*, pp. 36–46. London: SciTePress.

Strauch, S., V. Andrikopoulos, U. Breitenbücher, S. G. Sáez, O. Kopp, and F. Leymann (2013c). Using patterns to move the application data layer to the cloud. In *Proceedings of PATTERNS'13*, pp. 26–33. Melbourne: Xpert Publishing Services (XPS).

Taylor, I. J., E. Deelman, and D. B. Gannon (eds.) (2006). *Workflows for e-Science: Scientific Workflows for Grids*. New York: Springer.

Tran, V. T. K., K. Lee, A. Fekete, A. Liu, and J. Keung (2011). Size estimation of cloud migration projects with cloud migration point (CMP). In *Proceedings of ESEM'11*, pp. 265–274. New York: IEEE.

Varia, J. (2010). Migrating your existing applications to the AWS cloud. A phase-driven approach to cloud migration. *Amazon Web Services Blog.* http://aws.amazon .com/ blogs/aws/new-whitepaper-migrating-your-existing-applications-to-the-aws-cloud/.

Zinn, D., Q. Hart, B. Ludascher, and Y. Simmhan (2010). Streaming satellite data to cloud workflows for on-demand computing of environmental data products. In *Proceedings of WORKS'10*, pp. 1–8. New Orleans, LA: IEEE.

6

Closing the Gap between Cloud Providers and Scientific Users

David Susa, Harold Castro, and Mario Villamizar

CONTENTS

Summary

Cloud computing emerges as an alternative to traditional grid/cluster approaches. Particularly, software-as-a-service model can be an option to address the computational needs of small- and medium-size research groups, with little or no knowledge and resources to deal with the complexities of technology. Although there are still many problems to be solved and a long way to go before the solution is optimal, the e-Clouds project manages to hide the configuration required by public infrastructure-as-a-service (Iaas) providers by delivering ready-to-use scientific applications that take advantage of the cloud world.

6.1 Introduction

Everyday scientific work requires growing computational capacity to provide reliable and in-time results. The traditional approach to address these needs includes the acquisition, configuration, and maintenance of a large number of dedicated servers, introducing some constraints primarily associated with the elevated costs and complex information technology (IT) management. These high-performance platform requirements are a barrier to entry for small- and medium-size research groups.

Public cloud infrastructures present themselves as an alternative to traditional cluster and grid solutions [1]. Cloud providers offer a large set of infrastructure and application services to resemble the flexibility of private data centers, with the benefit of a pay-per-use model. This allows users to run a wide variety of applications, including enterprise, social, and mobile ones. The question is then: How to adapt this model for scientific requirements? As we show, almost everything required by a scientific application is available in the cloud. The main challenge is then that, despite the low prices and flexible set of resources, the complex deployment and execution procedures are an obstacle for researchers to adopt the technology.

This chapter describes the proposal developed under the e-Clouds project, which is designed to be a software-as-a-service (SaaS) marketplace for scientific applications running on top of a public cloud infrastructure. It will include a description of the most important aspects of e-Clouds architecture, emphasizing the different patterns applied for using cloud resources while hiding the complexity for the end user. A detailed presentation of the problems faced during development and testing of the first version is also included. A research group from the Alexander von Humboldt Institute for Biological Resources (Bogota, Columbia) is used as a case study for testing some of the ideas outlined and defining the future work for the project.

6.2 What Do Scientific Apps Require?

The concept of cloud computing was mainly developed by having in mind the usual enterprise web applications. Scientific applications have a particular set of requirements and characteristics. These new requirements demand a fundamentally different approach to problem solving. The next sections discuss the most important things to consider when relating science and computing.

6.2.1 Flexibility

There is a huge market of general-purpose scientific apps that cover the day-to-day tasks for the different disciplines. Despite that important offer, research work often forces technology to adapt, not the other way around. This is the reason why scientific applications need to be flexible regarding the kind of processing they support, the input data they receive, and the outputs they produce. Based on this, multiple file formats must be supported and a large number of configuration parameters become optional.

Flexibility poses a challenge when porting applications to a cloud platform, especially when offering them in an SaaS model. An attractive cloud proposal must then include at least the most common configuration options and a minimum degree of personalization. The way that this can be achieved can vary greatly between applications; this not only increases the overall complexity but also imposes some restrictions on the solution model.

6.2.2 Platform Maintenance

It is important to consider the wide offer of scientific apps and the multiple platforms in which they can run. This means that part of the migration to a cloud solution requires deciding under which configuration an application will run, including operating systems, compilers, and a set of external libraries, among others. This process becomes more complex when considering version management and personalized installations that are sometimes required.

6.2.3 High-Performance Computing

In a general definition, high performance in a scientific context means processing large data sets with large-scale resources. This imposes some challenges for the design of a cloud computing solution for researchers, including special attention to the details of the software and infrastructure offered by cloud providers.

Even though there is an important variety regarding hardware and software available, in general, cloud providers do not offer a platform designed specifically for scientific computation. A platform like this would require proper configuration of processing capability, high-throughput storage devices, and operating systems with optimized libraries for making calculus. Amazon Web Services (AWS) is actually making big efforts toward this by offering its EC2 (Elastic Compute Cloud) Cluster Compute and Cluster GPU instances. These instance types are specially designed for parallel applications that require a large amount of network communication. As their actual offer, cluster instances can be configured with up to 244 GB of RAM memory, 10 Gbps of input/output (I/O) performance, 88 processing units, and NVIDIA Tesla GPUs (graphics processing units) with "Fermi" architecture.

6.2.4 Data Communication

Although the available computing power is comparable to that found on grid/cluster infrastructures, cloud providers still have a long road to face to achieve the performance of these solutions. This seems to be especially true when talking about communications, which according to Jackson et al. [2] are the bottleneck for scientific cloud executions. Parallelization schemes often require data sharing between processes executing on different machines. Cloud infrastructure providers usually do not offer a dedicated data link or any guarantees regarding network throughput. This means that a scientific app running in a cloud has some limitations regarding the amount of data to communicate while maintaining the required performance.

6.2.5 Costs

Scale economy is the biggest driver for cloud computing. The low costs at which providers can acquire and maintain large data centers at geographically separate locations are the reason behind the success of the technology [3]. The idea of having access to thousands of servers just with a credit card and with no initial acquisition costs is simply amazing.

Small- and medium-size research groups almost always work with a small budget. An in-house infrastructure solution means that a great portion of the money that was destined to buy investigation equipment and finance their work now sits in a room full of servers that are not always used. Also, although some researchers are comfortable working with cluster configuration, parallelization, and computer programming, that is not the case for all of them. This lack of expertise means they have to pay a qualified professional to handle all the initial configuration and maintenance of their infrastructure.

6.2.6 Security and Reliability

One of the main concerns for these large-scale infrastructures is security. It is a common practice to share computing resources among scientists inside the same research group and with outsiders. Despite this shared environment, research work needs to remain confidential during its development, sometimes because of applicable legislation, until the scientist decides it is camera ready. Having this in mind, words like authorization, authentication, confidentiality, and accountability appear right away.

Together with a secure environment, scientific executions require a highly reliable platform. This is especially true when considering that some computations can take weeks or even months to complete. Losing a month of work just because of a server failure is simply not an option. This means that a platform for scientific executions must have adequate mechanisms to support these requirements.

6.3 Related Work

Approaches such as desktop grids and volunteer computing systems like BOINC [4], OurGrid [5], Integrate [6], and UnaGrid [7] have laid the bases to allow scientists to take advantage of large computing capabilities. Throughout these kinds of solutions, researchers are able to access high-performance platforms to run their workloads. However, the technical effort required to run a defined workload under such conditions is generally too high for an individual researcher with a tight schedule.

Recent developments in cloud computing solutions have aroused the interest of the scientific community. Much effort has been expended to achieve traditional cluster/grid performance in cloud environments. Some comparisons between cloud and grid have been made to show the benefits and challenges presented by both technologies [8–10], in some cases combining them through a hybrid approach [11]. Results from important research projects such as the Magellan report [12] have shown that cloud computing can fit scientific requirements under certain circumstances. Despite this, the technical complexity of the configuration process is still high.

Projects like the NGS Portal [13] have strived to integrate domain experts' knowledge into preconfigured application templates that are ready to run.

Some important developments for private cloud infrastructures and scientific workflow integration, such as Opal2 [14] and SciCumulus [15], have been made. Under this approach, a researcher is responsible for wrapping a scientific application in a preconfigured virtual machine (VM) or script. Although packaged VMs can be deployed automatically according to user requirements, administration problems arise right away when the number of supported applications increases and constant updates are necessary. This can be the case, for example, with the Scientific Computing as a Service (SCaaS) project [16].

There is also some work on an infrastructural level. Infrastructure-as-a-service (IaaS) solutions like OpenNebula [17, Eucalyptus [18], PiCloud [19], and Nimbus [20] offer a configuration environment especially designed for common scientific requirements. In this case, scientists who want to use these kinds of solutions need to be able to properly install and configure their own applications. There are also some upper-level commercial offerings like Cyclone [21] or SBGenomics [22], for which users can have access, in a SaaS model, to some commonly used applications like Hmmer [23], BLAST (Basic Local Alignment Search Tool) [24], or Gromacs [25]. These projects were built with some general needs in mind, making customization a complex process that depends on personal contact with the suppliers.

Different projects have focused on benchmarking conventional scientific solutions and workflows in both private and public cloud environments. Some studies have shown, for example, that a typical configuration in an IaaS provider like Amazon EC2 can be significantly slower than a modern high-performance system, especially when it comes to communication [2]. Despite this, it has been shown that research teams will adopt cloud computing over the next few years; in the meanwhile, cloud providers will likely improve their offering over important factors like costs, networking, administration, and elasticity [26].

Finally, it is worth mentioning some SaaS solutions that developed interesting models at an enterprise level. Among the most important ones, salesforce [27], ZOHO [28], and SuccessFactors [29] allow a wide variety of users to access complete business functionalities with low effort and at minimum costs. This way, small- and medium-size companies can benefit from solid solutions that fit their budget. The e-Clouds proposal is based on an integration of the ideas developed under some of these projects to meet the scientific requirements mentioned previously.

6.4 e-Clouds Architecture

6.4.1 Overview

e-Clouds is an effort to create an easy-to-use SaaS marketplace for scientific applications. As part of the initial proposal, the e-Clouds team will be in charge of supporting all the IT-related tasks, including designing and maintaining the platform execution. The target customers are researchers who will process their data in the e-Clouds platform. For the first version of the project, researchers will be able to store their data on e-Clouds and perform executions using a set of defined applications.

Initial tests of the proposal have been performed with research groups from the Humboldt Institute for Biological Resources [30]. Feedback received is being incorporated into a second version of the project, while looking for other research groups as early adopters. Throughout the following sections, some examples are shown by using a study case with a custom version of the Maxent software [31] for species habitat modeling. Maxent receives a file with the coordinates where the species actually live, and it generates a complete map that predicts alternative environments for this same species. Some ecosystem variables are fed into the system as map layers, together with a species definition to be processed. The output is a predictive model of the species geographic distribution by using a maximum entropy method. This process is commonly used in analytical biology mainly for conservation and species management.

The general architecture for the e-Clouds solution is presented in Figure 6.1. As presented, three basic components make up the proposal: the infrastructure provider (a public IaaS), the back end for jobs scheduling and control, and the front end that supports administration and user interaction. All the information regarding the users and their activities in the platform is stored inside a relational database. The communication between these components is made through a queuing service and the database records.

At first, a user registers in the e-Clouds web portal, pending approval. After an administrator approves the registration, the user will be able to access a private workspace through a username/password combination. When logged into the application, a user can manage his or her data (files and folders), launch and monitor the status of executions, and check his or her current account balance according to the costs of storage, computing, and communications.

A resource manager (RM) in the back end is in charge of controlling the cloud infrastructure according to the defined events or user actions. When a user launches, cancels, or modifies an execution, the web portal sends a request to the RM, which effectively takes the corresponding actions using the IaaS API. All actions inside the platform are stored as events in the database as they can have an impact on the execution total costs and serve for accountability.

The supporting platform for e-Clouds is based on three main elements: a standard Linux machine image, a reliable queuing service, and a scalable storage service. The Linux machine is specially crafted to execute certain boot steps when launched. Particularly, a machine, once started, is required to download the latest update of an agent program. This agent is in charge

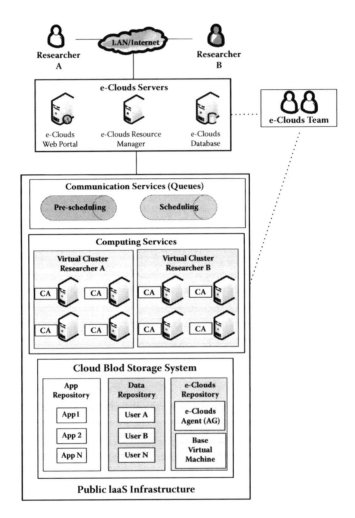

FIGURE 6.1
e-Clouds general architecture.

of executing and monitoring the jobs assigned to its particular machine inside the cluster.

6.4.2 e-Clouds Back End

6.4.2.1 Data Management

A key to achieving high performance in scientific workload executions is efficient management of data. For e-Clouds' particular case, it is possible to differentiate between four types of data: user files, transactional data, internode messages, and local data. The first two data types are required to be somehow persistent over time. The other two are associated with a defined execution, and they are only valid in the context of that execution. A description of the proposal to handle each data type is included in the following sections.

e-Clouds users can store, manage, and share their personal files within the platform. This includes the ability to organize them in a file system, with a directory structure, and generating publicly accessible URLs (uniform resource locators) for each one. To address these needs, the files reside in the storage service provided by the public IaaS, and the corresponding metadata are stored in a relational database. Every execution input and output file will be stored as part of the e-Clouds file system.

6.4.2.2 Resource Manager

Resource manager is the piece of software in charge of job scheduling and cluster control. It has the responsibility of starting and stopping new machines on demand, taking into account the pending jobs. It assigns a certain amount of workload to each machine in an effort to minimize the total costs and time. In addition to the management functions, the RM serves as a central communication channel between the front end and any machine in a cluster.

6.4.2.3 Queues for Asynchronous and Scalable Communications

Due to the dynamic and flexible scaling of the cloud infrastructure, communication between cluster nodes and other e-Clouds components is achieved using queues. These queues will be accessible from all components in the architecture and will offer a reliable and scalable asynchronous messaging system. An important benefit obtained with the use of queues is the possibility to buffer user jobs.

6.4.2.4 Agent in Processing Machines

The agent is a control program that resides in every machine that runs as part of an e-Clouds execution. It is in charge of managing the local job executions and communications of a machine with the RM. As part of its responsibilities, it handles app installation and configuration, launches assigned jobs, and monitors the overall execution progress, communicating any updates.

6.4.2.5 Billing

A billing system is included to provide information about the cost of the resources consumed. An event-based approach is taken to calculate resource usage for the current period. This means that the system is capable of tracking each event that somehow has an impact on the total costs and records the pertinent information. Metrics such as machine hours, data transfer to and from external sources, and the amount of stored information are part of the accounting process. The conversion logic between the IaaS provider and the e-Clouds pricing schema is also included.

6.4.3 e-Clouds Front End

6.4.3.1 For Researchers

Researchers and e-Clouds administrators can access a complete set of services via a web user interface. The web portal is the point of entry for users; it allows them to administer their personal workspace. In particular, registered users are able to submit and monitor jobs, upload and delete files and folders, check execution results, and track their periodic use of e-Clouds resources. The web portal aims to provide a simple tool for the e-Clouds user by hiding the underlying complexity of cloud administration.

6.4.3.2 For e-Clouds Administrators

The web portal mentioned also includes an administration panel for the e-Clouds team. Administrative users are able to manage users, stored files, and security permissions and check the event log to track the overall system activity. New applications are configured through the web administration panel by describing their basic characteristics, including inputs, outputs, and associated restrictions. The administration panel adapts dynamically to the application description and presents the relevant options to the end user.

6.5 e-Clouds Implementation

Throughout the next sections, the most important implementation details are presented. They cover each of the architectural decisions mentioned and include the particular technologies used.

6.5.1 e-Clouds Front End

The e-Clouds front end is designed to hide the underlying complexity of infrastructure configuration while allowing users to control the most important aspects of a scientific workload execution. To effectively achieve this, the web portal was built with the Ruby on Rails framework, which is certainly gaining popularity among developers. Traditional web application functionalities, including user management, file handling, and visual design, are built on top of some popular third-party libraries (gems). For example, the administration panel was developed using the Active Admin gem [32], which allows a fast buildup of dashboards and control features based on the model definition.

6.5.2 e-Clouds Back End

6.5.2.1 Data Management

6.5.2.1.1 User Files and Simple Storage Service Files Storage

Amazon's Simple Storage Service (S3) is, as its name implies, an easy-to-use web-accessible storage solution. S3 users can manage files with sizes ranging from 1 byte up to 5 TB through simple object access protocol (SOAP) and representational state transfer (REST) APIs. Downloading is possible through the hypertext transfer protocol (HTTP) and Bit Torrent protocol. The information stored in S3 will be replicated by default to AWS content distribution networks in multiple continents, with a guarantee of 99.999999999% durability.

As part of the e-Clouds initial version, users are able to store and manage their own files. These files can be used as inputs for a defined application execution or they can also be the results (outputs) of this same application execution. Having this in mind, a user should be able to handle his or her data much in the same way as with a local computer. This means the ability to organize files into folders and create, delete, download, and check their associated metadata.

AWS S3 service does not provide a complete directory structure that can be used to fulfill the requirements stated. This means that e-Clouds platform should provide an abstraction layer that allows a user to effectively manage data. As shown in Figure 6.2, a simple objects model relating files, directories, and users was created to solve this problem.

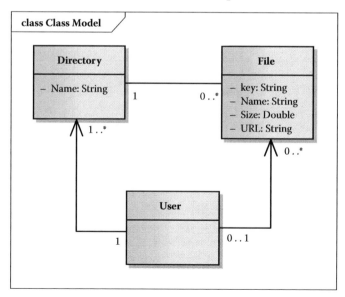

FIGURE 6.2
File system model.

As part of the model, directories are logical entities that provide file organization. Each directory can have files or subdirectories (children) associated. Files are stored in S3 inside defined buckets and folders. Each file object is aware of its physical location in S3 and also knows its parent directory. This way, the web application is capable of handling an organized file structure, and it is also possible to manage a first level of authorization by associating a file to a single user (its owner). In addition to the data model, a file browser set of views was necessary to facilitate user interaction.

6.5.2.1.2 Transactional Data

As part of the e-Clouds solution, there is a central database management system for storing data related to transactions. This includes but is not limited to basic data such as user profile, security associations, resources usage, S3 files metadata, applications, and of course, user executions. Besides application and execution information, the database contains what is expected to be in a standard web application. Database connections can only be established by the web portal and the RM to enhance security and make the administration (updates, tests, etc.) easier.

6.5.2.1.3 Local Storage

The main purpose of local storage is to store execution-related data in each cluster machine. It is primarily used as low-latency (and -cost) storage for installation files, libraries, input files, and execution results. All information that resides in local storage is considered ephemeral, so every time an execution finishes, output files and logs should be uploaded to S3 and indexed in the transactional data. Everything else that is on local storage will be erased once a machine shuts down.

6.5.2.2 Queue Messaging

Reliable message queues are the main communication channel between the different components that make up e-Clouds. At this first version, AWS Simple Queue Service (SQS) is used. Figure 6.3 shows how the information flows between the queues and the corresponding communicating entities. It is important to note that there are two main, always on, queues: prescheduling and scheduling queues. Also, there is one additional queue for each user execution, and it is used mainly for job assignment. It is created when execution is launched and destroyed when it finishes.

The prescheduling queue communicates messages that come from the web portal, to and from the RM. The scheduling queue has the initial messages that go from the RM to all the machines in a cluster and receives state updates from these same machines. At last, execution-specific queues are used to assign pending jobs to the associated machines.

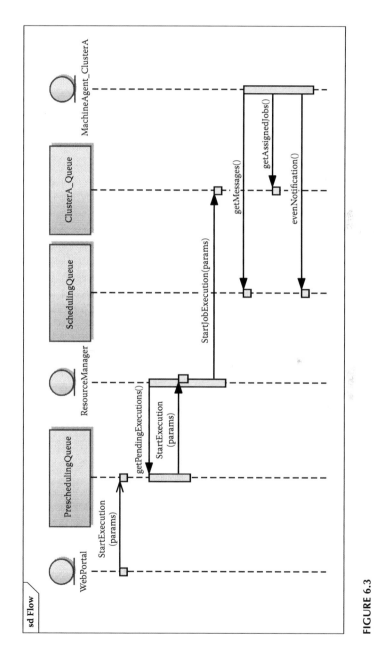

FIGURE 6.3
Queue communication sequence.

6.5.2.3 App Install and Configuration

Scientific application installation and configuration are the core back-end processes for the e-Clouds project. As the project's main objective is to support multiple heterogeneous apps for different disciplines, it is a challenge to establish a uniform general process for this. The proposed solution is based on Linux shell scripting and an object model including application commands, inputs, and parameters.

An execution starts with a standard x64 Debian-based Linux machine image. Depending on the application selected by the user, a specific script is downloaded and executed on a clean machine based on the image mentioned. This way, the complexity of multiple virtual images administration becomes a simple script management process, and e-Clouds is not charged by the cloud provider for storing preconfigured images.

6.5.2.3.1 Shell Scripts

As the deployment platform relies on a Linux-based image, shell commands are a simple choice for application installation and configuration. Each time a new app is uploaded, a shell script is required; this way, the installation process is automated, but the complexity of this process is separated from the other e-Clouds components. A basic checklist of what a script like this should have to be e-Clouds compatible would be the following:

Operating system and environment: This initial step involves configuring the requirements at the operating system level. This means creating the required users, files, folders, and so on; configuring the environment variables; and setting up general security.

External dependencies: This step installs the different application dependencies. This might include libraries, compilers, and other applications. It is important to consider the specific versions that are required so that the application works correctly.

Installation files: It is of course necessary to download and process all the installation files that make up the application itself. In some cases, this means downloading source code and compiling it each time. Again, it is important to consider version management to obtain the expected results.

Data files: These files can be considered as part of the installation files. In scientific applications, it is common practice to have large databases that contain information to be used within executions. Although these files' versions might change, they are usually static and common to all executions, so they cannot be set by the user as actual inputs.

In Figure 6.4, a sample script for installing the custom version of Maxent is shown. As shown, it uses the Ubuntu package manager (apt-get) to install some packages. Also, it processes some files using R language functions [33].

```
#!/bin/sh
sudo whoami
sudo echo 'deb http://cran.stat.ucla.edu/bin/linux/ubuntu jaunty/' >>/etc/apt/sources.list
#Os and Environment
sudo apt-get update
sudo apt-get -q -y install r-base r-base-dev
#External Dependencies
wget http://cran.r-project.org/src/contrib/Archive/sp/sp_0.9-99.tar.gz
sudo R CMD INSTALL sp_0.9-99.tar.gz
wget http://cran.r-project.org/src/contrib/Archive/raster/raster_1.9-67.tar.gz
sudo R CMD INSTALL raster_1.9-67.tar.gz
wget http://cran.r-project.org/src/contrib/Archive/dismo/dismo_0.7-17.tar.gz
sudo R CMD INSTALL dismo_0.7-17.tar.gz
wget http://cran.r-project.org/src/contrib/Archive/foreign/foreign_0.8-48.tar.gz
sudo R CMD INSTALL foreign_0.8-48.tar.gz
wget http://cran.r-project.org/src/contrib/Archive/lattice/lattice_0.20-0.tar.gza
sudo R CMD INSTALL lattice_0.20-0.tar.gz
wget http://cran.r-project.org/src/contrib/Archive/maptools/maptools_0.8-14.tar.gz
sudo R CMD INSTALL maptools_0.8-14.tar.gz
sudo apt-get -q -y install openjdk-7-jdk
sudo R CMD javareconf
wget http://cran.r-project.org/src/contrib/rJava_0.9-3.tar.gz
sudo R CMD INSTALL rJava_0.9-3.tar.gz
#Installation files
wget https://s3.amazonaws.com/../maxent.jar
sudo cp maxent.jar /usr/local/lib/R/site-library/dismo/java/
```

FIGURE 6.4
Custom Humboldt Maxent script.

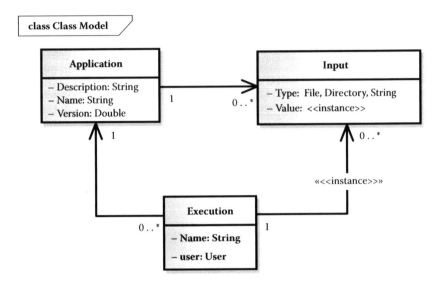

FIGURE 6.5
Application inputs and executions model.

6.5.2.3.2 Parameter Description

A simple model to describe application inputs was proposed as part of the
e-Clouds solution (Figure 6.5). In this model, an application can have many
inputs associated, and these inputs will be instantiated with a value for every
execution. Scientific applications can have basically two input types, a string
or a file. A third special input type is an e-Clouds directory so that all files
within that directory are considered inputs and multiple jobs are generated.

At the presentation layer, a user is able to assign input values for a specific
execution in three basic ways:

- Enter a string in a defined text field.
- Select a file from its workspace.
- Select a directory with at least one file from its workspace.

Once an execution is launched, application inputs are used to generate
an execution command by concatenating all the values set by the user and
the predefined ones, preceded by the appropriate prefixes. An example of
such a command can be found in Figure 6.6. As seen, each input file must

```
wget https://.../Atelopus_nicefori
wget https://.../full_map_layer_stack
R --no-save Atelopus_nicefori $HOME 'Atelopus_nicefori' full_map_layer_stack < Maxent2.R
```

FIGURE 6.6
Execution command example.

first be downloaded from its physical location in S3 so that it can be used locally by each machine in a cluster. The exact file name is part of the final execution command.

6.5.2.4 Scheduling

One of e-Clouds' main principles is to take advantage of the scientists' experience, especially when it comes to estimating the execution time for a certain workload. Based on this, the general scheduling process is as follows:

1. When a new application is configured to be part of e-Clouds, the uploader defines an estimation of the average execution time and suggested machine technical specifications. This will be the default configuration to run a scientific workload with that application. It is important to note that the uploading process must be accompanied by a domain expert with previous experience on the particular application.

2. Each time a researcher launches a new execution, the researcher is able to change the default values mentioned. This takes into account the researcher's knowledge regarding the amount of work he or she is sending. The value specified for the execution time is not a limitation, or a guarantee, of the real time the execution could take to finish. Also, the researcher is able to specify, based on the cloud offer, under which machine specifications he wants to run his own execution. This way, he is able to obtain from the e-Clouds platform a gross estimate of the execution's total cost.

3. Once an execution is launched, the scheduler takes into consideration the application inputs and the current configuration to decide the appropriate cluster configuration. In a general way, the basic decision process is shown in Figure 6.7.

The overall scheduling process is designed to optimize the resource use so that the total cost is minimized. To illustrate this, suppose an IaaS provider charges per hour or partial hour of computing use. An e-Clouds user starts an execution of a certain application with five different file inputs that are totally independent from each other. The current estimation indicates that the average processing time for each file will be 15 minutes. Considering this, the selected approach for task division is

- Launching two processing nodes.
- Processing files 1 to 3 in node A and files 4 and 5 in node B.
- Total processing time will be 45 minutes, which is the maximum between node A and node B total time.

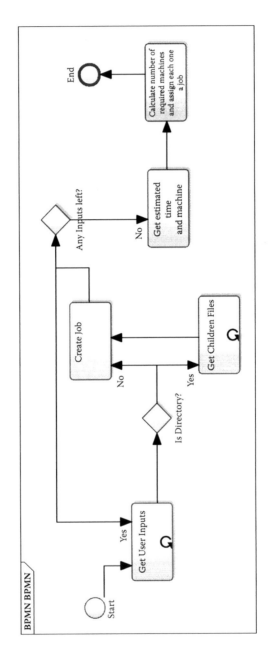

FIGURE 6.7
Scheduling initial process.

It is important to note that the total execution time could be reduced in this case by launching one machine per file, but that decision would imply a considerable cost increase.

6.5.3 e-Clouds IaaS Provider

One of the main goals for e-Clouds is to support multiple underlying infrastructure providers. This means that the user would be able to choose between different alternatives and select the one that best fits personal needs. Because there are no well-defined standards among cloud suppliers, this platform independence requires some extra development work from the e-Clouds team and is why this first version release supports just one of them. Taking all this into account and with the objective of launching a first version as soon as possible, the e-Clouds team selected AWS and Heroku for the initial testing.

Amazon Web Services has consolidated as one of the biggest and most complete public clouds offering IaaS. Its low prices and the flexible resource configuration are ideal for an initial testing of the e-Clouds ideas. Also, because of its large number of users and success cases, there is complete documentation available regarding performance benchmarking and testing under high-performance scenarios [34, 35].

Heroku is a cloud platform as a service (PaaS) offer for fast and simple deployment of web applications [36]. It actually supports Ruby, Node.js, Clojure, Java, Python, and Scala applications. As part of its proposal, Heroku has a free usage tier specially designed for developers. This free tier allows users to run testing environments in a basic configuration of one web server and an SQL database. Deployment can be easily done through a Git repository configuration and a set of command line tools.

As shown in Figure 6.8, e-Clouds web portal and RM are initially deployed in Heroku for ease. This includes the presentation layer developed under Ruby on Rails and the relational database running in PostgreSQL. On the other hand, AWS is used to run scientific workloads, store user files, and communicate information through reliable queues. A more detailed description of each one of these components can be found further in this chapter.

6.5.4 Monetization

As in any cloud solution, there is a cost transference between the infrastructure usage and the service delivered. During this initial phase, e-Clouds only charges each individual user by the different resources the user effectively consumed, including storage, computing hours, and communication. Although the total cost of an execution depends on these three basic factors, the charging model for each of them is slightly different.

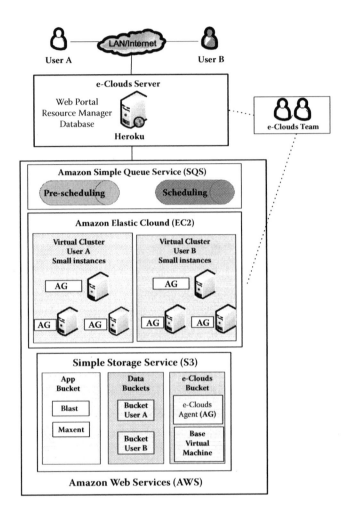

FIGURE 6.8
e-Clouds actual deployment.

6.5.4.1 Storage Cost

The user pays at the end of each month for the maximum amount of data that he stored during that period of time. For simplicity, the minimum charging unit is the gigabyte, so everything below that just rounds up. The formula for calculating total storage cost is then really simple:

Total storage cost = (Maximum # of gigabytes stored) * (Cost per gigabyte)

6.5.4.2 Computational Cost

Following some providers' trend (including AWS), the computational resources are charged on an hourly basis. This means that for every hour or partial hour that a machine is working on a particular execution, the user who launched that execution will have to pay a fixed amount. The hourly rate varies according to machine technical specifications, namely, more power means more money. Considering this, the formula for charging computational resource usage is the following:

Total computational cost = (Total # of machine hours per execution)
* (Hourly rate depending on machine size)

6.5.4.3 Additional Costs

The e-Clouds proposal includes an additional fixed rate for data communication and queue resource usage. This includes the cost for traffic to and from the machines and the costs associated with queue service. Every user should pay the same fixed rate for each new execution launched on e-Clouds. This way, the total additional cost would be calculated using the following formula:

Total additionals = (# of executions) * (Fixed rate for an execution)

There are two main reasons behind the decision for leaving these additional costs as a fixed rate. The first one is that, for most cloud providers, communication costs are almost insignificant compared to those of storage and computing. This is not true for all cases, but experience with some scientific applications showed that it somehow follows in general. The second reason for this management decision is that it can be complex to account for all communication processes that happen in a machine. This complexity has an impact not only on the e-Clouds daily operation but also on the development of new functionalities.

6.6 Results

Several tests were created to obtain an idea of the cost and time relationship under an AWS platform. Different instance types were used to check the performance under different technical scenarios. In particular, in AWS jargon, the following instance types were used: c1.medium (2 Cores 2,5 EC2 Units, 1.7 GB RAM, moderate I/O performance) and c1.xlarge (8 Cores 2,5 EC2 Units, 7 GB RAM). The instance selection resembles the machine specifications of a private cluster where some analyses were executed.

Maxent software is a simple .jar file that runs just like any other Java application. It displays a graphical interface with some options so that the user can define certain values that are relevant to the maximum entropy modeling approach. A custom version of Maxent was used to perform the tests. It was built using an R script, so that parameter configuration can be automatic when calling the original Maxent file, focusing on the special requirements of the Humboldt Institute and including some Java VM fine-tuning. This version is already configured as an e-Clouds application, accessible by all users. As shown in Figure 6.6, it depends on the packages dismo, maptools, sp, and rJava.

Three files are received by this application as parameters; the first is an input R script, which contains the R commands needed to analyze the data. The second is a stack file that contains different layers with characteristics of Colombia, such as temperature, humidity, altitude, and so on in a raw ".asc" data format. The third file contains the coordinates where a certain species has been spotted in Colombia, in a defined comma-separated value format. All the input files needed were previously uploaded to the S3-based e-Clouds file system under a user account. The outputs of the application differ based on the configuration, but usually include visual maps that show the resulting model for a particular species and can be exported to file formats (e.g., pdf or HTML).

Earlier, to execute the application, clusters were deployed in the university campus consisting of VMs using two cores of an Intel Core i7 processor and 8 GB of memory. In that execution, the files were stored in a network-attached storage. Similar jobs had been executed using the same input files used for the tests in e-Clouds. With these clusters, the average execution time for each job was 18 minutes.

As previously explained, the execution parameters are based on an initial time estimation made by the application configurator. The selection of these parameters affects other parameters, such as the total cost of the execution and the total time that it takes to finish. A user is capable of including the user's own estimation, based on the user's knowledge of the application and the data to be processed. The system recalculates the total costs and time when the parameters are changed.

Two different approaches were used: The first one seeks to minimize the total cost of the execution, and the other seeks to minimize the execution time. Previous estimations of the required time for a particular job execution to completion were made. The total execution time is calculated by multiplying the number of jobs by the expected time per job in minutes. Table 6.1 shows the results of the execution times and costs using different numbers of species.

The average installation time refers to the time spent on the application installation process. This process is only carried out once per machine and execution. The results show that the application install can be done on demand without significantly affecting the total time. It can also be seen that the times obtained from the earlier executions under private cluster environments are similar to the execution in AWS. It is important to note that, using a storage system like S3, it scales up adequately since the execution time is not affected by the number of machines.

TABLE 6.1

Initial Test Results

	Approach 1: Reduce Total Execution Time					Approach 2: Reduce Costs				
Number of Species	2	4	8	16	32	2	4	8	16	32
Number of VMs	1	2	2	4	8	1	2	2	4	8
VM type	c1.xlarge					c1.medium				
Cost per hour (US dollars)	0.66					0.17				
Average install time (min)	2.27	2.32	2.13	2.17	2.08	2.52	2.72	2.50	2.25	2.33
Average time per job (min)	10.50	10.85	10.59	10.08	10.10	15.50	15.65	14.30	14.10	14.65
Total execution time (min)	24.77	29.10	49.63	47.12	55.08	38.95	38.58	66.78	66.90	73.90
Used computing hours	1	2	2	4	8	1	2	3	6	10
Processing costs (US dollars)	0.66	1.32	1.32	2.64	5.28	0.17	0.33	0.50	0.99	1.65
Cost per species (US dollars)	0.33	0.33	0.17	0.17	0.17	0.08	0.08	0.06	0.06	0.05
Jobs per VM	2	2	4	4	4	2	2	4	4	4

Finally, it can be seen that the approach in which the total execution time was reduced has a significantly higher cost than the other one without reducing the total time in the same proportion. This means that it can be better to wait a little bit longer for an execution to complete, seeking to improve the final costs.

6.7 Conclusions and Future Work

Scientific cloud computing is still at an early age. Nevertheless, the academic community and commercial providers are making important efforts in this regard. New projects combining public clouds and traditional cluster/grid approaches will appear over the next few years. Of course, as cloud providers increase their capabilities to overcome the actual obstacles, new problems and challenges will appear. This is true not only from a technical perspective but also from an economical and cultural point of view.

A proposal for a scientific SaaS marketplace has been presented throughout the chapter. The most important architectural elements were described in addition to a brief overview of the work done so far in the e-Clouds project. The solution presented was based on the utilization of the resources provided by a public IaaS infrastructure, allowing small- and medium-size groups to access on demand ready-to-use applications while obtaining the benefit of the scale economies.

The work done so far covers the fundamental aspects of a solution of such a nature. The design decisions taken so far have aimed toward a functional and simple solution to the requirements mentioned at the beginning of this chapter. Additional work needs to be done to cover potential flaws and support scientific applications that are more complex. Further testing is required to adapt the solution to changing requirements and diverse research groups' needs.

Future plans for the e-Clouds project include the implementation of new features to favor collaboration among researchers and results validation [37]. This way, the platform can become part of scientific day-to-day work. In addition, new applications with different technical requirements will be tested, including large-scale and long-lasting executions. In this respect, there is some pending development regarding reliability and error handling.

Although the RM is capable of handling a minimum degree of parallelism, several improvements in both the front and back end need to be done to support the execution of highly parallel applications (using a message passing interface or graphics processing units) with effective resource management. Together with this, additional work is required to support application workflows transparently. Some already existing alternatives are being considered to support these requirements.

Finally, further optimization of resource scheduling is required to apply data-mining techniques to estimate execution time and cost and take advantage of the residual time of clusters and VMs. Although there is an important challenge in proposing a general solution, some opportunistic ideas are applicable to the e-Clouds scenario.

References

1. Rehr, J. J., F. D. Vila, J. P. Gardner, L. Svec, and M. Prange. Scientific Computing in the Cloud. *IEEE*, 12, no. 3 (2010): 34–43.
2. Jackson, K, et al. Performance analysis of high performance computing applications on the Amazon Web Services Cloud. *2010 IEEE Second International Conference on Cloud Computing Technology and Science (CloudCom)*, 2010: 159–168.
3. Wang, Lei, Jianfeng Zhan, Weisong Shi, and Yi Liang. In cloud, can scientific communities benefit from the economies of scale? *IEEE Transactions on Parallel and Distributed Systems* 23, no. 2 (2012): 296–303.
4. Anderson, David. BOINC: a system for public-resource computing and storage. In *Fifth IEEE/ACM International Workshop on Grid*, Pittsburgh: ACM, 2004, 4–10.
5. Andrade, Nazareno, Walfredo Cirne, Francisco Brasileiro, and Paulo Roisenberg. OurGrid: an approach to easily assemble grids with equitable resource sharing. *Lecture Notes in Computer Science* 2862/2003 (2003): 61–86.
6. Goldchleger, Andrei, Fabio Kon, Alfredo Goldman, Marcelo Finger, and Germano Capistrano Bezerra. InteGrade: object-oriented grid middleware leveraging the idle computing power of desktop machines. *Concurrency and Computation: Practice and Experience* 16, no. 5 (2004): 449–459.

7. Castro, H., Rosales, E., Villamizar, M., and Miller, A. UnaGrid—on demand opportunistic desktop grid. In *10th IEEE/ACM International Conference on Cluster, Cloud and Grid Computing*. Melbourne: IEEE, 2010, 661–666.

8. Foster, Ian, Yong Zhao, Ioan Raicu, and Shiyong Lu. Cloud computing and grid computing 360-degree compared. In *GCE '08 Grid Computing Environments Workshop*, 2008, 1–10.

9. Yu, Bing, Jing Tian, Shilong Ma, Shengwei Yi, and Dan Yu. Gird or cloud? Survey on scientific computing. In *2011 IEEE International Conference on Cloud Computing and Intelligence Systems (CCIS)*, 2011, 244–249.

10. Zhang, Shuai, Shufen Zhang, Xuebin Chen, and Xiuzhen Huo. The comparison between cloud computing and grid computing. In *2010 International Conference on Computer Application and System Modeling*, 2010, V11-72–V11-75.

11. Calatrava, Amanda, Germán Moltó, and Vicente Hernández. Combining grid and cloud resources for hybrid scientific computing executions. Presented at Third IEEE International Conference on Cloud Computing Technology and Science, 2011.

12. Office of Advanced Scientific Computing Research (ASCR). *The Magellan Report on Cloud Computing for Science*. Washington, DC: US Department of Energy, 2011.

13. National e-Infrastructure Service (NES). NGS Portal. n.d. http://www.ngs.ac.uk/use/tools/ngsportal (accessed June 4, 2013).

14. Krishnan, S., L. Clementi, Ren Jingyuan, P. Papadopoulos, and W. Li. Design and evaluation of Opal2: a toolkit for scientific software as a service. In *2009 World Conference on Services*. New York: IEEE, 2009, 709–716.

15. de Oliveira, Daniel, Fernanda Baião, and Marta Mattoso. SciCumulus: A lightweight cloud middleware to explore many task computing paradigm in scientific workflows. *2010 IEEE Third International Conference on Cloud Computing*, 2010, 378–385.

16. Prasad Saripalli, Curt Oldenburg, Ben Walters, and N. Radheshyam. Implementation and usability evaluation of a cloud platform for scientific computing as a service (SCaaS). In *Fourth IEEE International Conference on Utility and Cloud Computing (UCC)*, 2011, 345–354.

17. OpenNebula. OpenNebula. n.d. http://opennebula.org/about:contact.

18. Eucalyptus. Eucalyptus. n.d. http://www.eucalyptus.com/eucalyptus-cloud.

19. PiCloud. PiCloud. n.d. http://www.picloud.com/company/(accessed January 2, 2013).

20. Hoffa, C., et al. On the use of cloud computing for scientific workflows. In *IEEE Fourth International Conference on eScience (eScience 2008)*, Indianapolis, IN, 2008, 7–12.

21. Silicon Graphics International. January 2012. http://www.sgi.com/products/hpc_cloud/cyclone/.

22. Seven Bridges Genomics. Home page. n.d. https://www.sbgenomics.com (accessed June 4, 2013).

23. Howard Hughes Medical Institute. HMMER project. n.d. http://hmmer.janelia.org/about (accessed June 4, 2013).

24. National Library of Medicine. Blast Project. n.d. http://blast.ncbi.nlm.nih.gov/ (accessed June 4, 2013).

25. Gromacs. Home page. n.d. http://www.gromacs.org/.

26. Yang, Chia-Lee, Bang-Ning Hwang, and Benjamin J.C Yuan. Key consideration factors of adopting cloud computing for science. In *IEEE 4th International Conference on Cloud Computing Technology and Science*, 2012.

27. Fisher, Steve. The architecture of the Apex Platform, salesforce.com's platform for building on-demand applications. In *29th International Conference on Software Engineering—Companion*, 2007, 3.

28. Zoho Corporation. Home page. January 2012. http://www.zoho.com/.

29. SuccessFactors. Home page. January 2012. http://www.successfactors.com/.

30. Instituto Alexander Von Humboldt. Home page. January 2013. http://www.humboldt.org.co/iavh/.

31. Phillips, Steven J., Dudikc Miroslav, and Robert E. Schapire. Maxent software for species habitat modeling. In *Proceedings of the Twenty-First International Conference on Machine Learning*, 2004, 655–662.

32. VersaPay, Gregbell. ActiveAdmin. n.d. http://activeadmin.info/ (accessed June 4, 2013).

33. The R project. Home page. January 2013. http://www.r-project.org/.

34. Deelman, Ewa, et al. Data sharing options for scientific workflows on Amazon EC2. *SC10 Conference*, 2010.

35. Xiaoyong, Bai. High performance computing for finite element in cloud. In *2011 International Conference on Future Computer Sciences and Application*. New York: IEEE, 2011, 51–53.

36. Heroku. Home page. January 2013. http://www.heroku.com/.

37. Kumbhare, Alok Gautam, Yogesh Simmhan, and Viktor Prasanna. Designing a secure storage repository for sharing scientific datasets using public clouds. In *DataCloud-SC'11*, 2011.

7

Assembling Cloud-Based Geographic Information Systems: A Pragmatic Approach Using Off-the-Shelf Components

Muhammad Akmal, Ian Allison, and Horacio González–Vélez

CONTENTS

Summary

In this chapter, we present a novel systematic way of building a web-based geographic information system (GIS) running on cloud services. The proposed architecture aims to provide a design pattern for building a cloud-based GIS using simple and readily available low-cost tools with great overall system efficiency. The result of running the GIS using this paradigm is arguably reliable and available at low cost and with some platform independence. It has required significantly less time and effort to deploy when compared with standard cloud development. We present a case study based on road accidents using Microsoft Windows Azure and Amazon Web Services. In this case study, a GIS was created that helped in improvements of road conditions by identifying road accident hot spots in real time and on the real map. Later, authorities can use this information to implement preventive measures to reduce road accidents. This GIS can be implemented for any town, city, county, or region in the world as long as its satellite maps are available on Microsoft Bing Maps.

7.1 Introduction

In its canonical definition of cloud computing [10], the National Institute of Standards and Technology contended that "cloud computing is a model for enabling convenient, on-demand network access to a shared pool of configurable computing resources (e.g., networks, servers, storage, applications, and services) that can be rapidly provisioned and released with minimal management effort or service provider interaction."

Cloud computing is considered to be a new value-added paradigm for network computing, where higher efficiency, massive scalability, and speed rely on effective software development [1]. Having also capitalized on emerging business trends, such as capital asset control, carbon management, and total cost of ownership, its uniqueness lies in its simplicity: It has promised that every consumer, small business, and large organization will access any information technology (IT) platform as a utility [5].

Despite some initial security concerns and technical issues, an increasing number of businesses are considering moving their applications and services into "the Cloud."* Consequently, mainstream information communication technology (ICT) powerhouses such as Amazon, Microsoft, IBM, Apple, and Google are heavily investing in the provision and support of public cloud infrastructures. Although significant effort has been devoted to migrate generic web-based applications into the Cloud, scant research and development have been put into creating a generic design pattern for a geographic information system (GIS) pattern in the Cloud.

To address this need, this chapter presents a systematic model to develop and deploy cloud-enabled GIS applications based on a pattern-based architecture. The proposed architecture uses SQL Azure geospatial database, Microsoft Silverlight, Microsoft Bing Maps, .NET Framework 4, Windows Communication Foundation-Rich Internet Applications (WCF-RIA). Services and the resulting application have been fully deployed in two mainstream public cloud platforms, namely, Microsoft Windows Azure Platform and Amazon Web Services. It is therefore arguable that the lessons learned and

---

* In line with convention, we have capitalized "Cloud" when referring to the holistic global interconnected infrastructure as opposed to any specific infrastructure provided.

indeed the software components and techniques are applicable to the vast majority of public and private cloud infrastructures.

Contribution

By assembling software components on public cloud infrastructures, our approach is arguably extensible and open. Furthermore, as part of the recent trends of increasing reproducibility of software engineering contributions, we are publishing the entire software environments together with this chapter such that any cloud developer can use them.

7.2 Background

Incorporating geospatial and descriptive data, GISs are a holistic integration of hardware, software, and standardized formats for capturing/encoding, managing, analyzing, and displaying all forms of geographically referenced data [8]. GISs have long been used beyond the boundaries of geography, and they are typically an aggregation of nonhomogeneous architectural platforms, applications, and processing needs due to a heterogeneous universe of users in science, business, and society in general [16].

Formerly, organizations had to buy dedicated GIS software packages to use and manipulate data over their network. Currently, web-based GIS software packages are readily available, and many organizations use web-based GISs to increase their availability of information for public and internal use. However, one of the biggest problems with large GISs is that all data are not necessarily available from the start, and systems are commonly rolled out following geographic area patterns rather than system usage or resources.

Moreover, large GIS projects usually start from a small amount of data but expand rapidly as data increase, requiring expansion of installed resources. Ergo, system resources are not consumed in predictable patterns as different users may follow seasonal or incident-driven usage patterns. In time, as more data are collected, systems cover additional geographical areas, and this leads to the need to increase other resources required by the system. A canonical example is Google Maps. It can be seen that the street view is not available for every place in the world, but these views are growing with time as Google collects data.

The operations performed on geospatial data within a database require significant computational resources for processing, typically surpassing the standard departmental infrastructures of small- and medium-size enterprises. For example, the selection of locations (points) that reside inside a given region (polygon) is particularly computationally demanding as the region is typically represented by a significant polygon with thousands of points to be verified.

An increase in data size requires not only more storage but also other computational resources. For example, more data require more processing. As the area covered by the system and the number of people using the system increase, the capacity of the whole infrastructure must increase due to the inherent computational complexity. By providing elasticity and on-demand consumption, any feasible approach should allow the system to scale down when the user count is low and, conversely, should cope well with any usage peaks. Cloud computing ought to help overcome this scalability problem efficiently, with a service model that enables on-demand resource access by aggregating configurable computational resources that can be rapidly provisioned and released.

Although such a model seems to be tailored for data-driven environments such as a GIS, programming, manipulating, and processing geospatial data typically requires the inclusion of complex data structures and demanding mathematical transformations. Even though standard web programming techniques have evolved to be applied in cloud environments, only a few proven pattern-based programming paradigms have been successfully applied in the Cloud. So, it should be clear that seamless cloud deployment entails a substantial amount of work, and current GIS tools are typically associated with a GIS software developer while cloud ones can be locked to a given cloud provider. Arguably, to have different GISs in the Cloud there has to be an orchestration of infrastructure and applications that can show tangible financial and computing benefits.

Different authors have started to evaluate the distinct possibilities in this area. Some argued for the need for the next generation of cloud infrastructure to be supported through traditional multitier architectures [2], while others have pursued the provision of innovative object-oriented data models and algorithms to retrieve data in a distributed environment [19]. But, the vast majority of the relevant approaches encourage the creation of generic GIS web services on top of map image files and geographic information in general, accessed as a web service through the Google App Engine [3], MapReduce-BigTable [20], a private cloud [4], or simply the Internet [23]. Having reported initial performance figures on a par with similar server-based web services, the last two approaches are clearly not associated with public Cloud deployments, but all four are definitely representative of the growing trend for the provision of GISs as a service.

From a more general GIS perspective, recent works have advocated the creation of *spatial cloud computing*—a subarea in which the spatiotemporal principles of geoassembling cloud-based GIS spatial sciences, and by extension a well-designed GIS, can be effectively represented in the Cloud given the continuous nature of GIS constraints [11, 15, 21, 22]. As part of this growing trend, there have been few comparative analyses of GISs in grids and clouds [14], and on different public infrastructures [24], using ad hoc GIS deployments enabling performance-oriented environments. Nevertheless, the creation of standards and methodologies for the creation of web-enabled open-source systems in the geographical sciences remains an active research area [7, 17].

From an industrial perspective, Esri, a leading GIS vendor, has produced a case study for its own proprietary software without much to say on how to take advantage of platform as a service/software as a service (PaaS/SaaS) for building GIS in an open, generic way [6]. GIS Cloud Limited seems to be the only platform provider for building GIS in the cloud [9], but one can only build using their application programming interface (API) and supported tools and languages.

Arguably, additional research needs to be devoted to the efficient use of Cloud-enabled GISs to develop specific applications quickly using off-the-shelf components such as public maps and web services. Such a need has been recently highlighted by different authors as part of the emerging trends in interdisciplinary geographic processing on clouds [12, 13, 18].

RESEARCH GAP

It is therefore clear that there is a research gap in this area and the methodology proposed in the book chapter can be used to build GISs in the cloud, on either infrastructure as a service (IaaS) or PaaS using map services such as Bing Maps and Google Maps and deploying through the use of software patterns. The remainder of this chapter provides a systematic way to assemble a GIS in the cloud using public Cloud providers and off-the-shelf components that follow accepted best practices in software engineering.

7.3 Methodology

The key elements required to build a typical web-based GIS are as follows:

- a database management system (DBMS);
- base maps;
- a web server with some storage, high-speed network connection between the machines; and
- a secured Internet connection to provide service over the Internet.

Additional elements can be added, such as a mechanism to reduce failures (e.g., replication servers or storage disk mirrors), load-balancing systems, and a backup mechanism.

The DBMS ought to support geospatial data types, allow spatial indexes, and perform various operations on geospatial data using built-in functions.

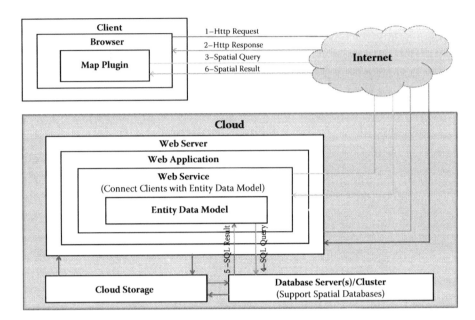

FIGURE 7.1
Block diagram for our cloud-based GIS architecture.

The web server should be able to support tools and APIs used to build the web application. For example, if the web application is built in .NET Framework, then an Apache-based web server might not work, and a web server based on the Internet Information Services (IIS) will be used.

Our cloud-based GIS architecture relies on the fact that cloud computing is an incremental approach to web-based systems. Figure 7.1 depicts the block diagram of the proposed architecture of a cloud-based GIS. It is obvious from the figure that the web server, database server, and storage will reside in the Cloud deployed as PaaS or IaaS. Each architectural component is explained next.

> **Client**: This is a computer/device with an Internet connection and a browser with the map plug-in (player) installed on it. For instance, if the map plug-in is developed using Microsoft Silverlight, then the Silverlight player needs to be installed on the client.
>
> **Map Plug-in**: A map plug-in is a specialized web-based map control that is capable of displaying base maps and supports displaying multiple layers of map objects, such as, polygons, locations, or lines. Since the idea of this project is to use Google or Microsoft Bing Maps as a base map because these are the most comprehensive maps available for the whole world, this could be a Google or Bing Maps control.

Web Server/Hosted Service: This is a web server in the Cloud with all necessary APIs installed on it. Depending on the cloud service provider and service provisioning model, this server differs from a "normal" web server. This cloud web server, ideally, must be horizontally and vertically scalable in the Cloud.

Storage Account: Deploys a cloud-based storage repository that will be used to store web applications.

Entity Framework Model: This creates a logical view of the database according to the business logic of the system and gives an abstract view of the database to the application, hence providing added security.

Web Services: A problem with rich Internet applications is coordinating the application logic between the middle tier and the presentation tier. Effective user experience requires the client to be aware of the application logic that resides on the server. But, it is cumbersome to develop and maintain the application logic on both the presentation tier and the middle tier. Our web service solves this problem by providing services that make the application logic on the server seamlessly available to the client. That is to say, it allows the client-side map plug-in to directly interact with the database in an easy, controlled, and secured way.

Database Server: The geospatial relational database of the application resides on this server. It has to be a powerful machine as most of the geospatial data manipulation is performed at this level. The actual DBMS can be Oracle, SQL server, or SQL Azure as all support geospatial databases.

Figure 7.1 also demonstrates that the process begins with the client when it sends an HTTP request for the application web page to the web server. The server responds and reads an appropriate web page from the storage, processes it, and then sends it to the client. Normally, this page will have a map control embedded in it that displays the base maps directly from Google or Bing.

Moreover, if a user interacts with the web page, for instance, by selecting a particular area on the map and searching for something within this area, the map control and the code behind generate a code-based geospatial query and send it to the web service that deals with database-related requests. The web service then passes it to the Entity Framework Model, which then converts the query into an SQL query tailored to the database server, which then processes it and sends back the results to the Entity Framework Model. The model finally invokes the web service communication with the client to display the relevant area on the map.

The key challenge is the *spatial data representation* between the client, the web service, and the database. The database can only keep spatial data in geometry or geography format, which is not directly produced by Google or Bing map

Cloud Computing with E-science Applications

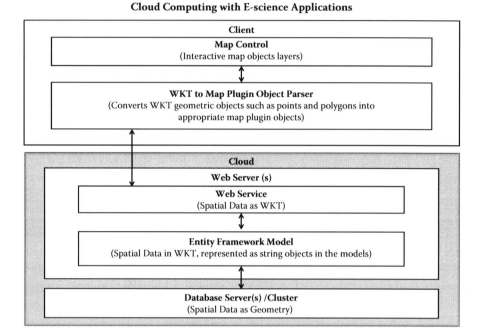

FIGURE 7.2
Data representation in the system. Note that our system exchanges data using the Open GIS Consortium's WKT (Well Known Text) format. (Redrawn from Yonggang Wang, Sheng Wang, and Daliang Zhou. In *Cloud Computing*, vol. 5931 of *Lecture Notes in Computer Science*, pages 322–331. Berlin: Springer-Verlag, 2009.)

controls. In addition, the Entity Framework Model in the web application does not typically support geometry or geography data types. To address this issue, there has to be a data representation or format for spatial data that can be easily exchanged between the DBMS geometry/geography data types, the map control objects, and the appropriate data types in the Entity Framework Model. The Open GIS Consortium's WKT (well-known text) and WKB (well-known binary) spatial data formats have been defined to enable such data exchange [19].

Figure 7.2 shows the representation of spatial data in different formats at different stages of the system, designed to deal with the data incompatibility problem. It is clear from the figure that the spatial data in the database will be held in the geometry format as it is easy to convert it into WKT format using built-in SQL functions.

The Entity Framework Model presents the spatial data in WKT format to the application and text, and it is stored as a string in the entities. The web service also deals with spatial data in WKT. One of the most important elements here is the WKT to the map object parser, which converts WKT data into appropriate map objects and vice versa. There is no direct way to convert a WKT geometry representation into map control objects. To overcome this problem, an API has been created that takes a WKT representation of geometry objects (polygons, line, point, etc.) and returns the equivalent appropriate map control objects. Moreover, the conversion of geometry data into WKT in the database may require use of views in the database, but this definitely compromises the performance and, if not done carefully, can cause the application to crash.

To prove the concept discussed in this methodology, a proof-of-concept (demo) application has been created in Microsoft Azure Cloud using a PaaS deployment. An image of this implementation has also been ported to the Amazon EC2 Cloud using an IaaS deployment. Although in this chapter we present full details related to the Azure implementation, it is highlighted that the Amazon deployment details are of a similar nature and are also introduced when relevant.

7.4 Implementation

Our application is a web-based GIS specifically designed to analyze the main causes of road accidents and dangerous road conditions in a specific region. This GIS application has been built in such a way that it can be seamlessly customized for any place in the world as long as its detailed maps are available on Bing. The main objectives of the demo application are to

- display a real map of the region where it will be used;
- provide a web-based graphical user interface to enable users to search road accidents in a selected region;
- interact with structured data for any part of the world where digital road maps are available;
- illustrate the reduction of effort for the implementation, management, and maintenance of cloud infrastructure; and
- demonstrate scalable and reliable behavior.

Titled the Road Accidents GIS, our demo application takes advantage of fundamental cloud computing capabilities such as scaling, redundancy, and reduced system management and administration employing different Microsoft technologies, such as Bing Maps, Silverlight, SQL Azure, WCF-RIA Services, and .NET. Figure 7.3 shows an overview of its implementation in Microsoft Azure.

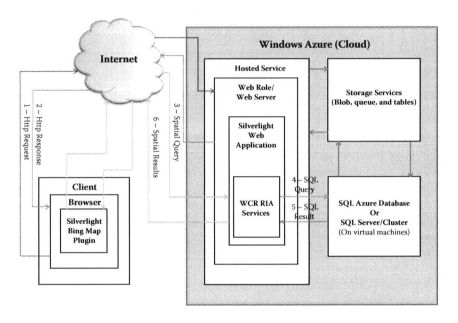

FIGURE 7.3
Architecture of our cloud-based GIS application in Azure.

It is noted that the key GIS components described in Section 7.3 have been instantiated under the standard Microsoft Azure platform as follows:

1. The map plug-in is deployed as a Silverlight plug-in, and a Microsoft Silverlight Bing Maps control has been used to display the maps.
2. The web server has been deployed as an Azure web role, which is actually a ready-made web server specially tuned for scaling. The web service has been distributed with the WCF-RIA services, which incorporate the business logic running in the web role (as the .NET Entity Framework Model) to the client.
3. SQL Azure has been used—instead of a separate database server—and only blob storage is used to store the application.

It is important to mention that all the images used in the interface need to be either embedded in the web application as a resource or kept in the Azure storage; their complete URLs used in the application are kept as image paths. Additional architectural components have been instantiated in the following manner:

Client: Computer or device with either MAC operating system (OS) X 10.5 or later running on an Intel machine or Windows XP or later OS. It must have a compatible browser with the Microsoft Silverlight plug-in installed.

Web Server/Hosted Service: This is a logical container that presents all machines (web/worker/virtual machine roles) running an application in Azure. The application can be elastically scaled up or down and accessed as a service hosted in Azure. In Azure, a hosted service has been taken as a web server that differs from a normal web server to constitute a PaaS infrastructure. In Amazon EC2, this is a machine instance with Apache or, potentially, IIS under an IaaS deployment.

Storage Account: The application uses the blob storage in Azure and the Elastic Block Storage in EC2.

Entity Framework Model and Web Services: WCF-RIA domain services provide data access to the Silverlight client according to the application logic running on the web server.

Database Server: SQL Azure has been used to support geospatial databases in Azure.

As described in Section 7.3, the client sends an HTTP request to the server. The server then processes it and sends a web page to the client, which then displays a Silverlight map on the screen as shown in Figure 7.4.

Specifically, the user selects a region type—which in turn requires region spatial data to be requested from the database and displayed on the map. To accomplish this, the client-side application generates a LINQ (Language

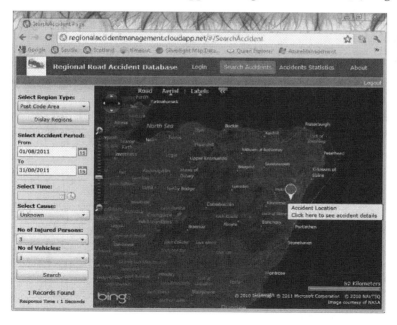

FIGURE 7.4
Main screen of our cloud-based GIS application in Azure.

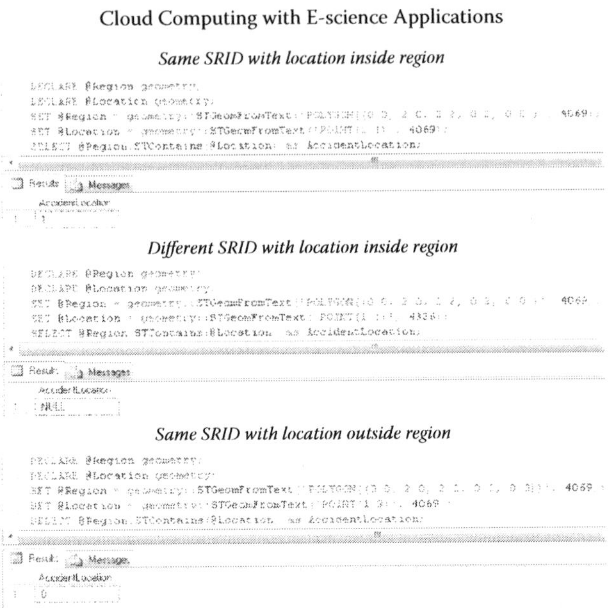

Cloud Computing with E-science Applications

Same SRID with location inside region

Different SRID with location inside region

Same SRID with location outside region

FIGURE 7.5
Example of s selection of points in a polygon.

Integrated [code-based] Query) and sends it to the WCF-RIA service, which then sends its equivalent SQL query to SQL Azure. The server then processes the query and sends back results to the WCF-RIA services, which then send enumerable entity objects to the client. Subsequently, Bing Maps object parser converts WKT region data into maps objects to be displayed on the map. Finally, the user selects the region and some accident search criteria to execute a search; as a result, the same process is performed to display the accident locations as pushpins on the map. The overall interaction is presented in Figure 7.4.

Note that, in this particular application, finding an accident in a region requires selection of points in a polygon. This is done using the STContains geometry function of SQL Azure, which determines if a geometry object, more specifically a spatial reference system identifier (SRID), is within a region. This function returns 1 if an object s, represented by an SRID, is inside a region and 0 otherwise, as shown in Figure 7.5. This functionality is particularly suitable for the Cloud as it enables demanding processing to be remotely commissioned.

7.4.1 Scaling and Fault Tolerance

Figure 7.6 represents a case in which three instances of our GIS application are running in the public Azure cloud. The three web roles represent three separate web servers running three copies of GIS applications

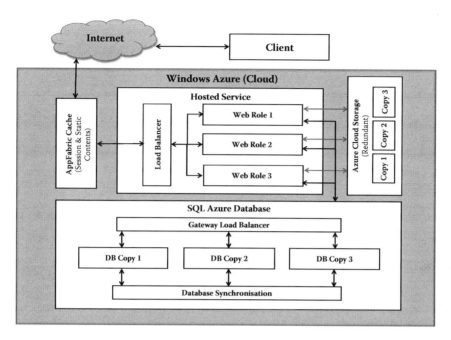

FIGURE 7.6
Implementation of cloud-based GIS application in Azure.

in the cloud. It is important to mention that each web role is an extra-small machine in Azure, which has a shared core, 768 MB memory, and 5 MB bandwidth. Azure allows measurement and monitoring of various performance parameters of the web roles such as the system load and, based on this information, scaling rules can be created. By using monitoring information, scaling rules, and Azure REST APIs, the number of web roles can be increased automatically (*horizontal scaling*). In some cases, if increasing the number of instances does not balance the load, the size of the web role machine can be increased (*vertical scaling*). Vertical scaling requires redeployment of the application, which is time consuming and may cause interruption to the service. Automatic scaling has not been implemented in our application.

Figure 7.6 shows a load balancer that evenly divides the workload among web roles, a common practice in any web-based architecture that poses a particular hurdle in Azure. Running multiple independent copies of the application on independent machines leads us to a problem if the session state variables are used in the application, as there is a possibility that different requests by the same client go to different web roles, as shown in Figure 7.7.

From this point, if an application is using session state variables, then sessions stored on one web role are not in the knowledge of other web roles. As there is even a possibility that a single client will be served by a

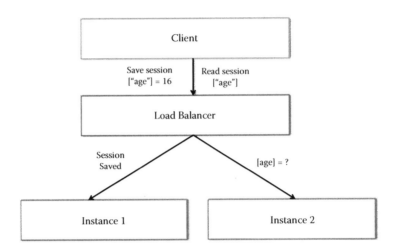

FIGURE 7.7
Session state problem with multiple web roles.

different web role for each request it makes, this will cause the application to malfunction. There are various ways to solve this problem but the quickest and easiest is to use the Azure AppFabric cache to hold the session state outside the web roles. Also, this cache can be used for in-memory buffering of static contents of the application, hence improving the performance.

As far as the hosted service is concerned, it is up to the certain subscriber how many web role instances the subscriber wants to create. Microsoft requires at least two web roles to ensure 99.9% uptime. In this case, three web role instances are running the application. In case of failure of a web role, other web roles take over and keep serving requests. As soon as Azure detects the failure, it replaces the faulty web role.

As Azure keeps by default three copies of storage and databases in three different domains of Microsoft data centers, there is no need to keep a backup database server or storage infrastructure. Once a fault is detected within a database, Azure automatically disconnects the faulty copy, acquires a new database instance from the Cloud, and synchronizes it with the remaining copies. As a result, when a failure occurs, it can have some effect on performance until the new database instance is ready. The same is true for storage.

7.5 Evaluation

For evaluation purposes, the applicative database has been populated with 150,000 demonstration road accident records for Scotland. To functionally test the system, a set of tasks has been devised to search the records in the

AB postcode area region corresponding to Aberdeen for each of the following 16 possibilities:

16 Search Possibilities: AB postcode

1. Date range
2. Date range and cause
3. Date range and time
4. Date range, cause, and time
5. Date range and number of persons
6. Date range, cause, and number of persons
7. Date range, time, and number of persons
8. Date range, time, number of persons, and cause
9. Date range and number of vehicles
10. Date range, cause, and number of vehicles
11. Date range, time, and number of vehicles
12. Date range, cause, time, and number of vehicles
13. Date range, number of persons, and number of vehicles
14. Date range, cause, number of persons, and number of vehicles
15. Date range, time, number of persons, and number of vehicles
16. Date range, cause, time, number of persons, and number of vehicles

We have selected such a narrow area and specific criteria because, as the search criteria become more specific, it decreases the number of accident locations in the result but requires more processing from the SQL Azure.

Then, each of these tasks has been carried out on the application with known selected field values, and the output was compared with previously known expected results. Any errors found were corrected in the code and then all the tests were performed repeatedly until all the results met the expected output.

7.5.1 Performance Testing

Performance testing is a key test as this application will arguably be running in a stressful environment and there will be a substantially large database for the system. In some unlikely cases when a user searches for longer date ranges, the result comprises thousands of records, potentially gigabytes of data. Then, since search results are downloaded from SQL Azure into the client, there is a possibility that the application can freeze or crash if the Internet speed is slow or the Internet connection is fluctuating between connected and disconnected states.

To perform stress testing, accidents in the AB postcode area have been searched for a 10-year range between January 1, 2001, and September 30, 2011, and without any other condition. This has put the application under extreme stress as in the normal case, producing 6,642 records for a single search (over 40% of the entire database). It was observed that when the number of resultant accident records was too high for a region (e.g., greater than 6,000), then on some occasions the application crashed and showed the error request time-out. After some research, it was realized that this was happening because of the default keep-alive time of the WCF-RIA service. For a WCF-RIA service request, the default keep-alive time on the client machine is 1 minute, and if a query response from the SQL server takes more than a minute, then the WCF-RIA service causes this error because the client side assumes that the connection with the domain service on the web server is broken.

After resolving this issue, this stress test has been performed repeatedly but from 10 different client machines at the same time. Although the response time was slow (up to 5 minutes), the overall architecture has worked, proving that the system is stable under stress conditions. For a small date range and with some conditions, the search is typically nimble, with response time within a couple of seconds.

Hence, our demo has worked in its most basic form, but there is a lot of room for improvement using the cloud computing capabilities. As an example, if a user requires extensive access to historical data (e.g., all the accidents that happened in the last 20 years), the result will be a region cluttered with thousands of accident location pins showing the problematic areas, which may be slightly difficult to read on a low-resolution screen.

7.5.2 Processing Time

A key performance parameter is how much time it takes to display a response/result on the client machine after sending a search request. This is dependent on the size of the accident search results, the client's Internet speed, processing required by the map objects parser, and processing required on SQL Azure. To analyze the processing time, a search is performed that returns records of all the accidents in a given postcode area (AB) in the database. This search actually executes a query in SQL Azure as shown in Listing 7.1.

Listing 7.1

SQL query in Azure

```
SELECT [RID] [AccID] [AccDate] [AccTime] [Cause] [Longitude]
[Latitude] [AccInjuredPersons] [AccVehicles]
FROM [AccidentManagement] [dbo] [viewPolyAccidents]
WHERE RegionID = AB
```

To check how much time SQL Azure takes to execute this query, the query has been executed 10 times via the SQL Server Management Studio. The average execution time was 76 seconds. When a similar search was performed

10 times on the same workstation via the accident application, it took variable times between 83 and 91 seconds, and the average time was 84 seconds. Here, it is clear that SQL Azure took most of the time, which means the rest of the processing in the application was fast, taking approximately 7 (84 − 77) seconds.

In both cases, the query returned 6,642 accident records, and the size of data returned by server was 543 KB. At that time, the application never crashed. It is important to mention that the Internet connection used was a 10-Mbps broadband connection with a DSL (digital subscriber line) test downloading speed of 3.84 Mbps and upload speed of 0.94 Mbps.

7.6 Conclusion and Future Work

This chapter addressed the need for generic design patterns for GIS cloud-enabled deployment. The proposed architecture utilizes the SQL Azure geospatial database in conjunction with other Microsoft technologies and services. This architecture is designed to be open and extensible.

Considering the original nature of the problem, it is noted that GIS deployment on clouds is per se a complex issue. It requires the seamless integration of distinct GIS capabilities to search, access, and utilize geospatial data with the cloud computing capabilities to configure, deploy, and manage computing infrastructure to permit the computability of intensive models and databases.

The chapter has provided a proposed methodology and architecture to enable systematic assembly of GISs in the cloud. This approach provides a viable solution to build stable and complex GISs in the cloud that can perform under extreme conditions, but there are a few minor issues, such as SQL server scaling and the need for a more robust and comprehensive API to convert WKT geospatial data into Bing, Google Maps, and other base maps objects. It is also interesting that an application built using this methodology can be implemented on IaaS and PaaS service models because it is implemented in PaaS and all the resources used in case of PaaS can be replicated in IaaS.

Further work should be carried out to evolve this paradigm for building GIS-oriented cost models for cloud computing where resources, computational and geographical, are correctly represented and priced. Additional areas to be tackled to further develop our concept are the following:

- Building a more comprehensive WKT geometry to map object parser APIs for all geometry types for different base maps, such as Google and Bing Maps.

- Currently, Bing Maps do not provide an object to represent multipolygons. We propose development of an API that allows developers to directly parse WKT multipolygons into a custom-built multipolygon object that can be directly displayed over Bing Maps Silverlight Control.

- We also propose further developing this concept using Google Maps and Java as Bing Maps does not have road maps for as many places as Google Maps.

- We propose further development of the Regional Accident Database project with geocoding—to find latitude and longitude from SRIDs—and capabilities such as location awareness to automatically select, as an example, all the accidents within a 5-mile radius of a selected location to be fed to a traffic warning system. This case study itself can be helpful to reduce road accidents.

- We propose added functionality to the system, such as having the system automatically analyze the road accident data and display a priority-based report that highlights the accident zones and tells us which zone needs the most attention.

Glossary

Client: This is a computer or device with an Internet connection and a browser with the map plug-in (player) installed on it. For instance, if the map plug-in is developed using Microsoft Silverlight, then the Silverlight player needs to be installed on the client.

The Cloud: Similar to Internet conventions, the capitalization of Cloud has been used when referring to the holistic global interconnected infrastructure as opposed to any specific generic infrastructure provided by a certain entity.

Database Server: Refers to the database services provided to the different software components. The geospatial relational database of the application resides on this server. It has to be underpinned by a powerful hardware configuration as most of the geospatial data manipulation will be performed here. In terms of software, the system can have Oracle, SQL server, or SQL Azure as all of these DBMSs support geospatial databases.

Entity Framework Model: Set of technologies in Microsoft .NET to develop data-oriented software applications. It allows the creation of a logical view of the database according to the business logic of the system and gives an abstracted view of the database to the application, hence providing added security.

Map Plug-in: A map plug-in is a specialized web-based map control that is capable of displaying base maps and supports a display of multiple layers of map objects, such as polygons, locations, or lines. Since the idea of this project is to use Google or Microsoft Bing maps as base maps because these are the most comprehensive maps available for the whole world, this could be a Google or Bing maps control.

SRID: The Spatial Reference System Identifier (SRID) contains standardized spatial coordinate system definitions for GIS.

Storage Account: Deploys a cloud-based storage repository that will be used to store web applications.

Web Server/Hosted Service: This is a web server in the cloud with all necessary APIs installed on it. Depending on the cloud services provider and service provisioning model, this server may differ from a normal web server. Ideally, this web server must be horizontally and vertically scalable in the cloud.

Web Services: Comprise software to enable the communication between devices on the web with an XML (extensible markup language) interface. A problem with rich Internet applications is coordinating application logic between the middle tier and the presentation tier. The best user experience requires the web services client to be aware of the application logic that resides on the server. In our case, the web service allows the client-side map plug-in to interact with the database in an easy, controlled, and secured way.

Acknowledgments

This work has been partly funded by the Horizon Fund for Universities of the Scottish Funding Council under the project Creating High-Value Cloud Services: Services to the Cloud (April 2011–March 2014).

References

1. Michael Armbrust, Armando Fox, Rean Griffith, Anthony D. Joseph, Randy H. Katz, Andy Konwinski, Gunho Lee, David A. Patterson, Ariel Rabkin, Ion Stoica, and Matei Zaharia. A view of cloud computing. *Communications of the ACM*, 53(4):50–58, 2010.
2. Muzafar Ahmad Bhat, Razeef Mohd Shah, and Bashir Ahmad. Cloud computing: a solution to geographical information systems (GIS). *International Journal on Computer Science and Engineering*, 3(2):594–600, February 2011.

3. J. D. Blower. GIS in the cloud: implementing a web map service on Google App Engine. In Lindi Liao, editor, *Proceedings of the 1st International Conference and Exhibition on Computing for Geospatial Research & Application, COM.Geo '10*, pages 1–4. Washington: ACM, 2010.

4. Claudius M. Bürger, Stefan Kollet, Jens Schumacher, and Detlef Bösel. Introduction of a web service for cloud computing with the integrated hydrologic simulation platform ParFlow. *Computers & Geosciences*, 48:334–336, 2012.

5. Rajkumar Buyya, Chee Shin Yeo, Srikumar Venugopal, James Broberg, and Ivona Brandic. Cloud computing and emerging IT platforms: vision, hype, and reality for delivering computing as the 5th utility. *Future Generation Computer Systems*, 25(6):599–616, 2009.

6. David Chappell. *GIS in the Cloud*. White paper. San Francisco: Chappel and Associates, September 2010.

7. Daoyi Chen, Shahriar Shams, César Carmona-Moreno, and Andrea Leone. Assessment of open source GIS software for water resources management in developing countries. *Journal of Hydro-environment Research*, 4(3):253–264, 2010.

8. Paul J. Curran. Geographic information systems. *Area*, 16(2):153–158, 1984.

9. GIS Cloud. Home page. http://www.giscloud.com/ (accessed December 1, 2012).

10. Timothy Granc and Peter Mell. *The NIST Definition of Cloud Computing*. Special Publication 800-145. Gaithersburg, MD: Information Technology Laboratory, National Institute of Standards and Technology, September 2011.

11. Qunying Huang, Zhenlong Li, Jizhe Xia, Yunfeng Jiang, Chen Xu, Kai Liu, Manzhu Yu, and Chaowei Yang. Accelerating geocomputation with cloud computing. In Xuan Shi, Volodymyr Kindratenko, and Chaowei Yang, editors, *Modern Accelerator Technologies for Geographic Information Science*, pages 41–51. New York: Springer-Verlag, 2013.

12. Hassan A. Karimi and Duangduen Roongpiboonsopit. Are clouds ready for geoprocessing? In Ivan Ivanov, Marten van Sinderen, and Boris Shishkov, editors, *Cloud Computing and Services Science*, Service Science: Research and Innovations in the Service Economy, pages 295–312. New York: Springer, 2012.

13. Zaheer Khan, David Ludlow, Richard McClatchey, and Ashiq Anjum. An architecture for integrated intelligence in urban management using cloud computing. *Journal of Cloud Computing*, 1(1):1–14, 2012.

14. Ick-Hoi Kim and Ming-Hsiang Tsou. Enabling digital earth simulation models using cloud computing or grid computing—two approaches supporting high-performance GIS simulation frameworks. *International Journal of Digital Earth*, 6(4):383–403, 2013.

15. Zhihui Liu. Typical characteristics of cloud GIS and several key issues of cloud spatial decision support system. In M. Surendra Prasad Babu, editor, *Proceedings of the 4th International Conference on Software Engineering and Service Science, ICSESS*, pages 668–671, Beijing: IEEE, 2013.

16. L. D. Murphy. Geographic information systems: are they decision support systems? In *Proceedings of the 28th Hawaii International Conference on System Sciences, HICSS '95*, vol. 4, pages 131–140. Maui: IEEE, 1995.

17. Markus Neteler, M. Hamish Bowman, Martin Landa, and Markus Metz. GRASS GIS: a multi-purpose open source GIS. *Environmental Modelling & Software*, 31:124–130, 2012.

18. Wang Tao. Interdisciplinary urban GIS for smart cities: advancements and opportunities. *Geo-spatial Information Science*, 16(1):25–34, 2013.

19. Yonggang Wang, Sheng Wang, and Daliang Zhou. Retrieving and index-ing spatial data in the cloud computing environment. In Martin Gilje Jaatun, Gansen Zhao, and Chunming Rong, editors, *Proceedings of the First International Conference on Cloud Computing, CloudCom 2009*, vol. 5931 of *Lecture Notes in Computer Science*, pages 322–331. Berlin: Springer-Verlag, 2009.

20. Yang Xiaoqiang and Deng Yuejin. Exploration of cloud computing technolo-gies for geographic information services. In Yu Liu and Aijun Chen, editors, *Proceedings of the 18th International Conference on Geoinformatics 2010*, pages 1–5. Beijing: IEEE, 2010.

21. Chaowei Yang, Michael Goodchild, Qunying Huang, Doug Nebert, Robert Raskin, Yan Xu, Myra Bambacus, and Daniel Fay. Spatial cloud computing: how can the geospatial sciences use and help shape cloud computing? *International Journal of Digital Earth*, 4(4):305–329, 2011.

22. Chaowei Yang and Qunying Huang. *Spatial Cloud Computing: A Practical Approach*. Boca Raton, FL: CRC Press, 2013.

23. Chaowei Yang, David W. Wong, Ruixin Yang, Menas Kafatos, and Qi Li. Performance-improving techniques in web-based GIS. *International Journal of Geographical Information Science*, 19(3):319–342, 2005.

24. Peng Yue, Hongxiu Zhou, Jianya Gong, and Lei Hu. Geoprocessing in cloud computing platforms—a comparative analysis. *International Journal of Digital Earth*, 6(4):1–22, 2012.

8

HCloud, a Healthcare-Oriented
Cloud System with Improved Efficiency
in Biomedical Data Processing

Ye Li, Chenguang He, Xiaomao Fan, Xucan Huang, and Yunpeng Cai

CONTENTS

Summary

As an emerging state-of-the-art technology, cloud computing has been applied to an extensive range of real-life situations. Health care service is one of such important application fields. We developed a ubiquitous health care system, named HCloud, after comprehensive evaluation of requirements of health care applications. It is provided based on a cloud computing platform with characteristics of loose coupling algorithm modules and powerful parallel computing capabilities that compute the details of those indicators for the purpose of preventive health care service. First, raw physiological signals are collected from the body sensors by wired or wireless connections and then transmitted through a gateway to the cloud platform, where storage and analysis of the health status are performed through data-mining technologies. Last, results and suggestions can be fed back to the users instantly for implementing personalized services that are delivered via a heterogeneous network. The proposed system can support huge physiological data storage; process heterogeneous data for various health care applications, such as automated electrocardiogram (ECG) analysis; and provide an early warning mechanism for chronic diseases. The architecture of the HCloud platform for physiological data storage, computing, data mining, and feature selections is described. Also, an online analysis scheme combined with a Map-Reduce parallel framework is designed to improve the platform's capabilities. Performance evaluation based on testing and experiments under various conditions have demonstrated the effectiveness and usability of this system.

8.1 Introduction

As the pace of life grows ever faster these days, the physical and psychological pressures on people are increasing ceaselessly, which raises the potential risks for many chronic diseases, such as high blood pressure (HBP), diabetes, and coronary disease. The large proportion of other adults who are suffering from "subhealthy" status (also called "the third state," which is between health and disease) are mainly engaged in brain work under high mental pressure. A total of 75% of the world's population are jeopardized by this negative situation [1]. The majority of them are white-collar workers and social elites, and they pay increasing attention to their health while hoping to obtain preventive health examinations periodically. Particularly, the aging issue worldwide is becoming more serious, and we need measures to improve the quality of life and launch chronic disease surveillance for elderly people. However, it is well known that public medical resources are usually insufficient and imbalanced in geographical distribution. According to a report of the World Health Organization (WHO), in recent years the regional average in Africa of those who can use an improved sanitation facility is about only 34% of the population; it is 94% in Europe [2]. So, to meet the needs of subhealthy groups, the aging population, and other people who need such services, a health care system that can provide self-monitoring of healthy status, provide early warning of disease, and even deliver analysis reports instantly is proposed and the option is becoming increasingly popular.

A health care system is a smart information system that can provide people with some basic health monitoring and physiological index analysis services. It is hard to share with isolated professional medical services such as PACs (picture archiving and communication systems), EHRs (electronic health records), and HISs (hospital information systems) without Internet-based technologies. Not long ago, this kind of system usually was implemented with a traditional MIS (management information system) mode, which is not capable of implementing sufficient health care services on a uniform platform, even though it may exploit several isolated Internet technologies. Currently, cloud computing, as an emerging state-of-the-art information technology (IT) platform, can provide economical and on-demand services for customers. It provides characteristics of high performance and transparent features to end users that can fulfill the flexibility and scalability of service-oriented systems. Such a system can meet the infrastructure demand for the health care system. With the rapid progress of cloud capacity, increasing applications and services are provided as anything as a service (XaaS) mode (e.g., security as a service, testing as a service, database as a service, and even everything as a service) [3]. Google Docs, Amazon S3 (Simple Storage Service), Ping Identity, and Microsoft Azure are popular products for online office application service, storage service, security service, and private platform service, respectively. Linthicum [4] investigated the services-oriented architecture (SOA) techniques applied in enterprise application integration (EAI) and the refining of the National Institute of Science and Technology (NIST) models with the XaaS concept. Another alternative cloud model is the Jericho Cloud Cube [5], which focuses on the collaboration-oriented architecture (COA) to ensure secure business collaboration in deperimeterized environments. The Distributed Management Task Force (DMTF) proposes a cloud architecture [6] that consists of a set of interfaces with specific definitions. Samba develops logical data models (LDMs) [7] for analyzing cloud architectural requirements to facilitate traceability between business requirements and cloud architecture implementations.

In our work, we propose a cloud-based system for preventive health care, named HCloud [8, 9], which implements both the analysis of physiological signal data and the early warning mechanisms for diseases. Unlike previous works, we take advantage of cloud storage for the large number of multimodal physiological signal data with heterogeneous characteristics. Implementations of cloud storage for physiological data, as well as computing for data mining and feature selections, are presented here. Performance evaluations based on the testing demonstrated the effectiveness and usability of the system.

This remainder of the chapter is organized as follows: The application scenarios, architecture, and key components of HCloud are described in Section 8.2, which also provides the details of the data analysis services in HCloud. Section 8.3 gives the details of the Map-Reduce paradigm immersed in the platform, as well as the health care services that HCloud can provide. Section 8.4 provides information on performance testing and evaluation; a conclusion is drawn in Section 8.5.

8.2 The HCloud Platform

In recent years, researchers have made some useful attempts to implement an efficient health care system with the power of cloud computing. For example, Zhang et al. [10] proposed a cloud security model based on EHRs that belongs to an MIS. Narayanan et al. [11] discussed access control to the health care system by considering role task management. Chang et al. [12] proposed an ecosystem approach to solve patient-centric health care and evidence-based medicine. However, previous works mainly focused on the storage, access, and management of private health information, which are quite primitive applications regardless of the computational power of the cloud platforms. It is expected that a cloud-based system not only stores the information but also performs basic analysis of health status and provides useful advice or warnings to patients, which is the purpose of our work.

8.2.1 Challenges to the Cloud Platform for Health Care

Cloud computing inherited the features of high-performance parallel computing, distributed computing, and grid computing and further developed these techniques to achieve location transparency to the end user and improve user experiences. In addition, a general cloud platform must face some challenges in health care service areas, as discussed next.

8.2.1.1 Heterogeneous Physiological Data Access

One challenging task for the health care cloud system is to handle the multimodal and nonstationary characteristics of special physiological signals, such as those for HBP, electrocardiography (ECG), and photoplethysmography (PPG). It is quite an inefficient job for a cloud system to store the numeric small-size physiological signal data on the ordinary distributed file system. Most of the distributed file systems are more suitable for large-size file storage than for small-size storage because there are bottlenecks for small-size files to access metadata on the local file system, which may result in synchronous problems (e.g., on the GlusterFS [13]). So, physiological signals with various formats should be well managed and processed to provide efficient instant services to individuals, and a consistent storage system is needed to adapt to this situation.

8.2.1.2 Multiscale File Storage and Integration

There are mainly two categories of application data in the system. One is trivial and has a small amount of data for temporal signals, such as the aforementioned physiological signal data; the other has huge-scale graphic data generated by drawing server clusters. Those large numbers of semistructured or unstructured health records, as well as massive trivial files, are all adapted to NoSQL [14] (not only in SQL) databases instead of traditional relational ones, which has natural advantages for easily expanding horizontally. The key idea of NoSQL is that it employs a loosely coupled data model and has neither a fixed table schema nor joint operations. Hence, it is appropriate for the high-performance requirement when accessing large files, especially for those without fixed structure.

8.2.1.3 Adaptive Algorithms for Different Targets

In consideration of the various characteristics of body health and aspects of a monitoring system, we need different algorithms for different signal processing and data mining. Note that the analysis routines should be easily configurable and adaptive to concurrent requests from users. Therefore, a flexible algorithm scheme should be developed for the sake of on-demand services. To cope with the irregularity of the data structure, a self-defined message head would be utilized to identify the call of various routines. Moreover, many real-time tasks should be addressed, and high concurrent mechanisms will be the major concern, although a general cloud may not need to face too many real-time transactions.

8.2.1.4 Visualization of Health Analysis

Another challenging task is the visualization of analysis results, which are usually computationally intensive with a large amount of graphic data for drawing. Careful considerations should be taken into account for efficient storage of and access to the huge graphic data generated by the analysis results. The system should be capable of handling the different types of data visualization adaptively. These fundamental characteristics are very different from the features of grid computing, which aimed at special applications and were difficult to operate for unprofessional users (e.g., in scientific exploration projects).

8.2.1.5 Data Convergence of Biosensors and Cloud

The method of service delivery is another important factor that affects the usability of the system. The user of the health cloud system should be provided with some easy-to-use data collectors, which is unlike general cloud users, who are only concerned about their data in the cloud. Necessary facilities should also be equipped with a friendly interface to transmit data to the cloud. For instance, a mobile phone is an appropriate front-end device [15] and always acts as a gateway into the system. However, seamless data fusion from signals collected for data processing in the cloud should be a concern.

8.2.2 Architecture of the HCloud Platform

The HCloud will face thousands of potential customers, including physicians and home users who care about their own health status. After analysis of the requirements of the application, an entire information flow of the HCloud analysis procedures is depicted as Figure 8.1.

The workflow of the proposed platform comprises three main steps. First, raw physiological signals are collected from the body sensors by wired or wireless means. Then, they are transmitted to the cloud platform to store and

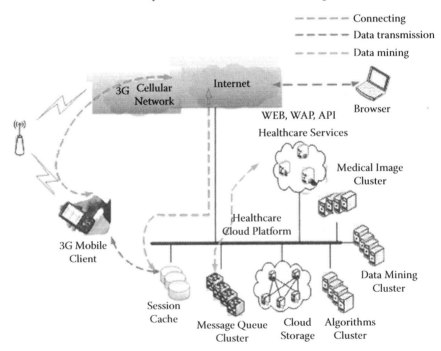

FIGURE 8.1
Overview of HCloud system.

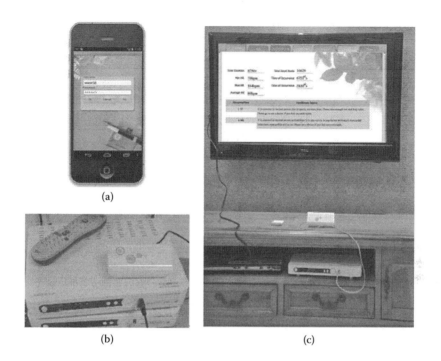

(a)

(b)   (c)

FIGURE 8.2
Mobile, set-top box, and home use scenario.

analyze the health status by means of data-mining technology. Finally, results can be fed back instantly and suggestions made to the users. Meanwhile, physicians in the community or a hospital can obtain their patient's health information from the Internet and provide some suggestions to the patients on daily dietary, exercise, and medication needs. If a patient's physiological index is abnormal after testing, the health care cloud system would give out warning information to the patient as well as send the message to their family members and physicians.

Also, there are two main activities of users: uploading the raw data and browsing the diagnosis results. Uploading activities are divided into two steps: The first is to transmit physiological data, which are collected by sensors, to the gateway through short-distance transmission by wired USB or wireless Bluetooth means. The second is to deliver the data from the gateway to the cloud servers through Internet protocol (IP) networks. A set-top box (for USB) or mobile phone (for Bluetooth) plays the important role of the gateway to relay data. In HCloud, apart from the mobile phone, a TV set is another friendly interface for the aging population. So, with the assistance of a set-top box, services would be delivered through a TV cable ubiquitously. Figures 8.2a and 8.2b show a mobile phone and a top-set box as gateways for data transmission, respectively, and Figure 8.2c is a scenario for home use.

8.2.3 Functionalities of Components in HCloud

As mentioned, HCloud is planned in the near future to serve hundreds of city-dwelling families via Internet health monitoring and will support thousands of potential customers, including physicians and whoever cares about their health. So, a higher number of concurrent transactions and shorter response time are requirements of our system, as is large-scale graphic drawing for visualized results, which need to be stored and accessed efficiently to provide instant services to individuals.

Therefore, the six-layer architecture of the private HCloud platform is proposed with the philosophy of inheriting the software as a service (SaaS) of the NIST model and introducing the in-source/outsource concept into development. Each layer's content and function are interpreted as service interaction, service presentation, session cache, cloud engine, medical data mining, and cloud storage, respectively, as shown in Figure 8.3.

- *Service interaction* is a top layer; users can interoperate with the terminals, such as 3G mobile phone, set-top box, and browser on a computer, to collect and upload original physiological data as well as download analysis results from the cloud servers.

- *Service presentation* can be regarded as an interface with various kinds of services, such as the wireless application protocol (WAP), web, or image provider. A load balance mechanism is introduced to the system on this layer so that the web/WAP servers can work cooperatively as a cluster to maintain optimal performance. The server cluster shares users' requests to the website together to meet the high concurrence requirement of health care services and to better ensure quality of service (QoS). In addition, some runtime information is reserved at this level.

- The *session cache* stores the user's sessions on one hand, which maintains the authentication and certification status the first time when the user accesses the services (i.e., information of service interaction through the presentation layer). To share the sessions among all web/WAP servers, this platform uses a separate session server to save the session data, which can solve the problem of session status sharing when there is load balancing. On the other hand, a memory cache is adopted to expedite access to the results data. Cache servers are specialized servers used to save users' pages, documents, profiles, and so on temporarily. This kind of server can reduce the capacity of network exchange because displaying graphs of physiological signals would take a long time. This platform will draw the images when users access the service the first time and save the graphs in the graph servers while registering the session keys in the cache server. If the users want to browse a previously generated image, the image will be loaded from the graph servers directly

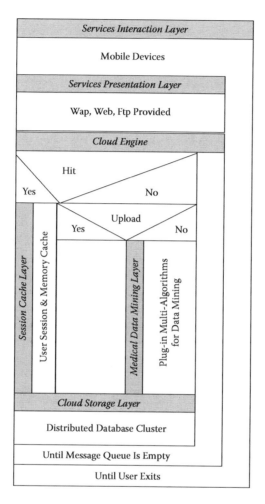

FIGURE 8.3
Six layers of a health care cloud platform. (Redrawn from He, C., Fan, X., and Li, Y., *IEEE Transactions on Biomedical Engineering* 60, no. 1 (January 2013): 230–234.)

without redrawing. In fact, this session's architecture is a big hashed key/value table.

- A *cloud engine* is a dispatcher to make components cooperate with each other to make the cloud run using a message-driven mode via a message queue (MQ) cluster. The functionalities of the queue management are provided by this layer, which is regarded as a critical core for scheduling tasks. A message is the unit of data transmitted between two modules. The MQ is a container to save the message during the transmission process. By introducing a message queue into the platform, the coupling between different modules can be significantly reduced.

- *Medical data mining* is a cluster of algorithms, including data preprocessing, data analysis, mining algorithms, and visualization processing. This layer can handle the data transmitted from the front end and generate the results back to the database. Other algorithms can also be easily plugged in if needed. These data-mining clusters are made up of servers executing data-mining algorithms. These algorithms can process the raw physiological signal data transmitted from the front end and generate the resulting data to write back to the cloud storage, launching a message into MQ middleware to indicate the subscriber to deal with. The tasks executed by the mining servers include data preprocessing, analysis, mining model tasks, and medical image drawing.

- *Cloud storage* provides data resources for the entire health cloud platform, including user information, vital signs, health records, and graphic data for processing. Physiological data collected from body area networks and massive graphic data for distributed processing of such data-intensive tasks are the primary contents. The cloud storage organizes various types of storage devices together by network and provides data storage and business access for outside applications, with the aid of cluster applications and grid or distributed file system technologies. A service sequence diagram of the HCloud can be represented as in Figure 8.4.

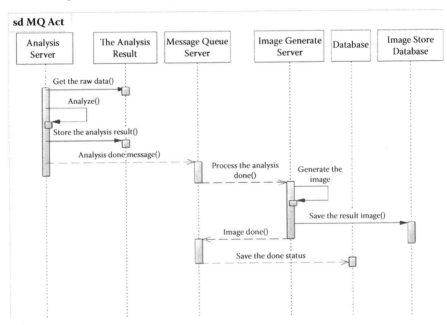

FIGURE 8.4
Service sequence map of HCloud.

8.2.4 Key Components Implementation in HCloud

There are three key components addressed here:

- The *MQ* is a kind of component that provides an asynchronous communication protocol to achieve independence between the message sender and the receiver. The queue is utilized for storing the event message generated by a sender (publisher), and all of the listeners (subscribers) who are interested in this kind of event can fetch this message to process a predefined routine. All communication parties should follow the same protocol (e.g., advanced MQ protocol) and utilize an available MQ API [16] to operate the queue. It is not necessary that two parties must know each other exists. After a message is inserted into a queue, it will be preserved and not deleted from the MQ until the corresponding subscriber reconnects to the system. Messages can be exchanged on the process, application, or even inter-cloud level. The queue resides in the cloud just as an engine to drive the system. In addition, different message heads identify the analysis algorithm and indicate the subsequent behaviors of the cloud.

- *The plug-in algorithm framework* is developed with respect to the extendability of various services, which is based on the publish/subscribe mechanism to provide customized functions conveniently, not only for health care but also for other services. The whole system can reduce the module coupling by adopting this algorithm. For instance, various data-mining algorithms, such as analysis of peripheral vascular function, instantaneous heart rate, chaotic characteristics of the power spectral density, and so on, should be adopted to perform automatically according to different analyzed signals. Every different function can subscribe to the different themes of the message, which is classified by a message head. In other words, using such Publish/Subscribe mechanisms, different mining functions can be called on by listening to the corresponding types of message the function is interested in. Accordingly, we designed an abstract core class named the *CoreStubClass*, including a private attribute *analysisKind* and an abstract method *handleRequest(int)*. This class communicates with *Message* via MQ, and the other kinds of concrete implementation classes extended the *CoreInterface* (e.g., *SignalFilterCore*, *ECGAnalysisCore* and other data-mining classes). Figure 8.5 shows the class framework of the plug-in algorithm.

- *Distributed storage* is the basis of cloud storage. The structural model of cloud storage is composed of four layers: the storage layer, the platform management layer, the application interface, and the access layer, as shown in Figure 8.6.

The core layer is the platform management layer, which ensures the reliable storage and efficient access of the large amounts of semistructured or

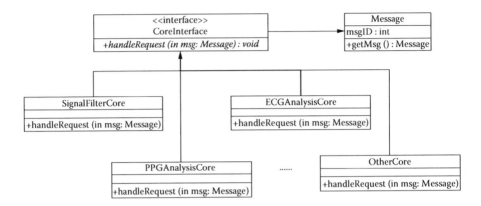

FIGURE 8.5
Plug-in algorithm framework. (Redrawn from He, C., Fan, X., and Li, Y., *IEEE Transactions on Biomedical Engineering* 60, no. 1 (January 2013): 230–234.)

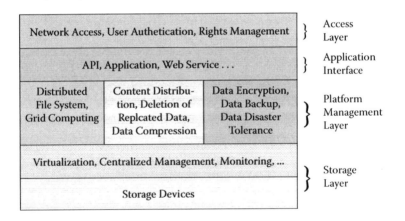

FIGURE 8.6
Four-layer concept model of cloud storage. (Redrawn from Fan, X., He, C., Cai, Y. and Li, Y., in *IEEE CloudCom 2012*, Taipei, December 3–6, 2012, 705–710.)

unstructured health documents, as well as miscellaneous signal files, with the power of NoSQL databases and distributed file systems. A NoSQL database is easy to integrate into distributed file systems. For instance, the Hadoop Distributed File System (HDFS) and Google's Cluster File System (GlusterFS) all have friendly interfaces to NoSQL databases such as Cassandra, MongoDB, and HBase. Moreover, sharding is another major characteristic of this distributed database to gain increased availability. Redundancy among these pieces of shards and different views of the same data provide consistency to a large extent. This mechanism can guarantee the integration of global data and transparency to users. For large and distributed storage, this architecture provides more convenience for data retrieval with better scalability as well as stability and persistency.

TABLE 8.1

Algorithm of *MQonMapReduce*

```
Input: iSize,oSize
Output: Boolean
  BEGIN
    initial a thread
    while (message from MQ is Null)
      if (message is DONE) return true;
        // DONE is a particular message from MQ
      else
        loop;
      load Mapper(message)
        // parallel Map function for data processing and
        analysis according to different message
      iSize←sizeof (files of input directory)
        //compute the total size of files under input directory
      if iSize>SIZE
        MQonMapReduce(iSize, oSize);
          //immerge trivial input files to a larger one
      else {
        oSize←sizeof(output directory)
          //compute the total size of files under output directory
        if oSize>SIZE
          MQonMapReduce(iSize, oSize);
            //immerge trivial output files to a larger one
        else
          send a message DONE to MQ;
            // process of all data are accomplished; }
  END
```

A *parallel computing framework* based on the Map-Reduce framework should be combined with the data stored in the NoSQL database (e.g., MongoDB) to deliver complex analytics, and data processing for such physiological data processing is always bound to the CPU (central processing unit). The map function can be designed to handle part of the data, while the reduce function is to merge the output produced by the map function and then output all of the filtered results. According to our six-layer HCloud, MQ can be utilized as a scheduler to cooperate with the Map-Reduce scheme. Generally, data analysis flow on this scheme can be described as in Table 8.1. Please note that this algorithm is a recursive procedure. The constant *SIZE* is a threshold according to the different scales of physiological data processing, which indicates the Map-Reduce procedure is to merge with the size of data scale. It will not be finished until all files are generated by the threshold of SIZE. Hadoop and the Map-Reduce programming paradigm already have a substantial base in the bioinformatics community [17] (e.g., monitoring of long-term ECG for individuals). The next section introduces the details of the ECG data process with this paradigm.

8.3 Provision of Health Care Information

The physiological indices of a person may show abnormality when his or her health status is trapped in a bad state. Hence, it is necessary for the user to obtain an *early warning* of this health status. HCloud can provide further semiautomatic or automatic analysis of physiological data by means of statistics and pattern recognition as well as data-mining methods. This section introduces the online analyses of ECG with the Map-Reduce scheme and presentation of the results via the cloud platform, as well as presentation of the other physiological signals (e.g., of the PPG and HBP), which can provide convenient, customized health care service. An actual mobile health system is described in Figure 8.7.

8.3.1 Online Analysis of ECG Data

The ECG is a transthoracic (across the thorax or chest) interpretation of the electrical activity of the heart over a period of time, as detected by electrodes attached to the outer surface of the skin and recorded by a device external to the body [18, 19]. It is utilized to measure the rate and regularity of heartbeats as well as the size and position of the chambers, which can diagnose atrial premature beats (APBs), arrhythmia, myocardial ischemia, and so on. Apart from simple records and general instructions, the system also provides detailed ECG physiological indexes for medical experts who need to obtain complete user information for diagnosis. The most common forms of arrhythmia, such as bigeminy, premature beats, bradycardia, and the frequency of occurrence are autoanalyzed by related algorithms.

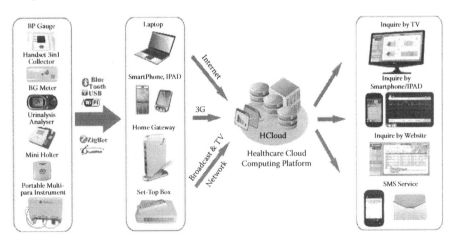

FIGURE 8.7
Architecture for an actual mobile health system.

FIGURE 8.8
Four phases of ECG data processing.

8.3.1.1 ECG Data Features

ECG data is collected from the human body with the frequency range between 0.05 and 100 Hz, and its amplitude is only several millivolts. Hence, disturbances from the environment always must be a concern, to avoid baseline wander of signals, and then QRS wave detection will be performed after signal denoising. The ECG data process can be divided into four phases: filtering baseline wander, denoising, detecting the QRS wave, and other postprocess phases, as shown in Figure 8.8.

For the long-term (24-hour) ECG data, data can be up to 12 MB at a sampling frequency of 150 Hz. An integer coefficient digital filter transfer formula specific to the ECG signals is as follows.

$$H(z) = z^{-87} - \left(\frac{1}{30} \times \frac{1-z^{-90}}{1-z^{-3}} \right)^2 \tag{8.1}$$

Then, the filtering baseline wander iteration function will be derived from Equation (8.1):

$$y_1(n) = x(n) - x(n-90) + y_1(n-3)$$

$$y_2(n) = y_1(n) - y_1(n-90) + y_2(n-3) \tag{8.2}$$

$$y(n) = x(n-87) - \frac{y_2(n)}{900}$$

where $x(n)$ is the original input signal, and $y(n)$ is the output signal (i.e., filtering baseline wander data). After testing the execution speed of the serial program with the raw data, the results showed that filtering the baseline wander phase took about 70% of the entire ECG processing time in our existing HCloud platform [20], which was the bottleneck of our ECG analysis algorithm, as Table 8.2 shows. If a single machine or serial programming is merely adopted, the user experience would be worse. Therefore, it is considered to be the first part of computing in parallel on the Map-Reduce framework.

8.3.1.2 Parallel Programming of the ECG Data Process

According to the analysis presented, the computing overload of the ECG data process is mainly at the phase of filtering the baseline wander. So, filtering the baseline wander should be parallelized first. Raw data from the

TABLE 8.2

Running Time of Each Part in Single Machine (Bold Indicates Longest)

Raw Data/ Measuring Time (hours)	Runtime (seconds)				
	Filtering Baseline Wander	Denoising	Detecting the QRS Wave	Others	Total Time
00007732/21.5	**194.281000**	7.829000	59.406000	179.515000	441.046000
00016412/13	**117.110000**	4.734000	11.734000	5.079000	138.657000
0039720/3	**26.672000**	1.078000	1.469000	2.656000	31.875000
01297217/12	**104.000000**	4.344000	6.500000	32.156000	147.015000
01334816/14.5	**124.797000**	5.234000	12.594000	27.094000	169.719000

client should be uploaded to MongoDB server by the remote user. To obtain these input data by the demon running on the Hadoop platform, they should first be downloaded from MongoDB locally and then uploaded to HDFS for further analysis and processing. In this scenario, data are pulled from MongoDB and processed within Hadoop via one or more Map-Reduce jobs. Output from these Map-Reduce jobs can then be written back to MongoDB for later querying and ad hoc analysis. Communication between client and platform is implemented by RabbitMQ, which is a popular MQ middleware. Assuming that the data contain 24-hour ECG signals, you might consider designing three map functions, each for 8 hours of the data, and then computing in parallel. Replications of processed data on HDFS are output files. The whole procedure is shown in Figure 8.9.

Each split represents a segment of data in the filtering baseline wander parallel programming. Each phase has key value pairs as input and output, in which the key stands for the data fragmentation flag, while the value stands for the ECG raw data or processed data (i.e., in the form of *<split-flag, ECG raw data>*, and *<flag, 24-hours processed signal>*, respectively). Since the default implementation of the interface *InputFormat* in Hadoop is *TextInputFormat*,

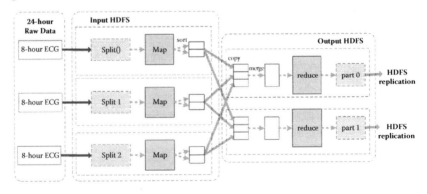

FIGURE 8.9
The Map-Reduce for filtering baseline wander.

which is not suitable for our application (ECG data are in byte format), we override *InputFormat* and define a format whose keys are represented by *TextWritable* and the values are the file contents, represented by *BytesWritable*.

Figure 8.10 shows a case of process results or original data and output data on HDFS.

8.3.1.3 Summary Report of ECG Status

After analysis of ECG data, the graphic-processing clusters can generate the graphic data, and the visualization of results would be provided to the end user. Many can be stored in the cluster for later requests, and all temporary pictures can be drawn at the client so that the computational load on the server decreases. All in all, the HCloud system can store large heterogeneous ECG data and compute the corresponding physiological indices within 5 seconds. If one of the physiological indices becomes abnormal, it generates disease warnings on time and sends messages to physicians. At the same time, it mines the deep regularity or characters from the long-term historical ECG data, which can help physicians better monitor an individual's health state. The system also provides detailed ECG physiological indices for medical experts, including normal-to-normal (NN) intervals, standard deviation of NN intervals (SDNN), standard deviation of the averages of NN intervals in all 5-minute segments of the entire recording (SDANN), heart rate variability (HRV) triangular indices, the triangular interpolation of the NN interval histogram (TINN), and so on.

Figure 8.11 is a simple summary text report of an ECG in the health care cloud system and shows the measurement duration, in how many seconds the abnormal ECG also can be obtained, as Figure 8.12a shows. On the other hand, HCloud generated the Poincaré image of one's heart movement while measuring ECG signals each time to represent the heart's chaos characteristics, as shown in Figure 8.12b.

Through spectrum analysis, we can draw a conclusion that everyone's heart movement has a chaos status. Chaos features represent the health status of the heart. They can illustrate the capability of the heart to adapt to different situations or body conditions at different times. The image shape is always heart-like if one is in good health. The more irregular the heart movement is, the more dangerous life is.

8.3.2 Other Physiological Diagnostic Data

Other important physiological information should be manually set into the system to build a traceable case history for health status (e.g., medical inspection data as an output of a blood glucose meter or urine analyzer from authorities, e.g., a qualified administration, a professional institution, a hospital, etc.). A follow-up survey of physiological data with diagnostic value is also presented. For instance, the general body information and health knowledge database are established and further instructions are given to customers. These services are optional to the end users.

```
bit&r410-220:/usr/src/hadoop-1.0.4$ hadoop fs -ls /user/huangxuecan
Found 20 items
drwxr-xr-x  - bit supergroup  0 2013-08-13 16:50 /user/huangxuecan/87064624_56518049_0668FF54_20130813_084931
drwxr-xr-x  - bit supergroup  0 2013-08-13 16:51 /user/huangxuecan/87064624_56518049_0668FF54_20130813_084931_out
drwxr-xr-x  - bit supergroup  0 2013-08-13 16:51 /user/huangxuecan/87064624_56518049_0668FF54_20130813_084947
drwxr-xr-x  - bit supergroup  0 2013-08-13 16:51 /user/huangxuecan/87064624_56518049_0668FF54_20130813_084947_out
drwxr-xr-x  - bit supergroup  0 2013-08-13 16:53 /user/huangxuecan/87064624_56518049_0668FF54_20130813_085204
drwxr-xr-x  - bit supergroup  0 2013-08-13 16:53 /user/huangxuecan/87064624_56518049_0668FF54_20130813_085204_out
drwxr-xr-x  - bit supergroup  0 2013-08-13 16:53 /user/huangxuecan/87064624_56518049_0668FF54_20130813_085218
drwxr-xr-x  - bit supergroup  0 2013-08-13 16:53 /user/huangxuecan/87064624_56518049_0668FF54_20130813_085218_out
drwxr-xr-x  - bit supergroup  0 2013-08-13 17:06 /user/huangxuecan/87064624_56518049_0668FF54_20130813_090545
drwxr-xr-x  - bit supergroup  0 2013-08-13 17:07 /user/huangxuecan/87064624_56518049_0668FF54_20130813_090545_out
drwxr-xr-x  - bit supergroup  0 2013-08-13 17:07 /user/huangxuecan/87064624_56518049_0668FF54_20130813_090641
drwxr-xr-x  - bit supergroup  0 2013-08-13 17:08 /user/huangxuecan/87064624_56518049_0668FF54_20130813_090641_out
drwxr-xr-x  - bit supergroup  0 2013-08-13 17:08 /user/huangxuecan/87064624_56518049_0668FF54_20130813_090732
drwxr-xr-x  - bit supergroup  0 2013-08-13 17:08 /user/huangxuecan/87064624_56518049_0668FF54_20130813_090732_out
drwxr-xr-x  - bit supergroup  0 2013-08-13 17:09 /user/huangxuecan/87064624_56518049_0668FF54_20130813_090835
drwxr-xr-x  - bit supergroup  0 2013-08-13 17:09 /user/huangxuecan/87064624_56518049_0668FF54_20130813_090835_out
drwxr-xr-x  - bit supergroup  0 2013-08-13 17:10 /user/huangxuecan/87064624_56518049_0668FF54_20130813_090848
drwxr-xr-x  - bit supergroup  0 2013-08-13 17:10 /user/huangxuecan/87064624_56518049_0668FF54_20130813_090848_out
drwxr-xr-x  - bit supergroup  0 2013-08-13 17:18 /user/huangxuecan/87064624_56518049_0668FF54_20130813_091711
drwxr-xr-x  - bit supergroup  0 2013-08-13 17:18 /user/huangxuecan/87064624_56518049_0668FF54_20130813_091711_out
```

FIGURE 8.10
Original input data and output data after processing.

From 2013 07 To 2013 08

Index	min	Time to min	MAX	Time to MAX	AVG	Normal
SDNN	21.06	2013-08-03 16:58:41	65.85	2013-08-03 13:36:54	52.88	17-79ms
HRVI	5.20	2013-08-03 16:58:41	12.27	2013-07-31 14:32:00	9.02	20-52
PNN50	5.35	2013-08-03 16:50:52	25.75	2013-08-03 13:36:54	17.97	8-16
RMSSD	13.45	2013-08-03 16:58:41	60.66	2013-08-03 17:09:34	40.59	15-39ms
TP	247.22	2013-08-03 16:58:41	2164.25	2013-07-31 14:32:00	1476.34	2448-4484ms²
LF	0.00	2013-07-31 14:53:00	870.12	2013-07-31 14:32:00	554.41	754-1586ms²
HF	14.61	2013-08-03 16:58:41	456.06	2013-08-01 15:50:55	254.11	772-1178ms²
LF/HF	2.06	2013-07-31 14:32:00	11.50	2013-08-03 17:03:43	4.33	1.5-2.0

Overall Report: HRV is normal (For proffessinal only)

Health Advices: 1. Do not be so fussy, balance the meal

2. Do exercise often, keep your weight

3. Do loose yourself, keep positive

4. Do not drink much, no smoking

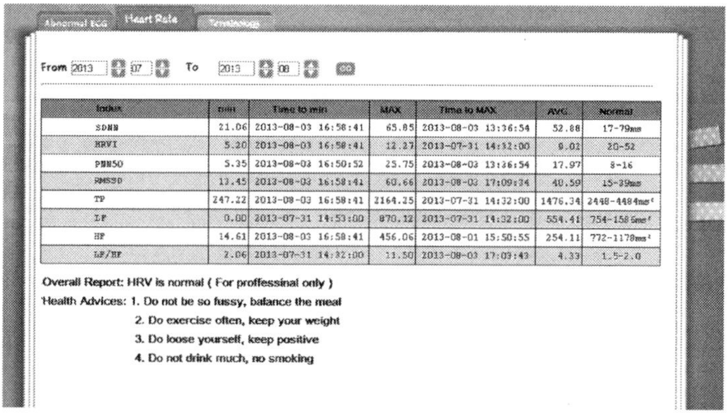

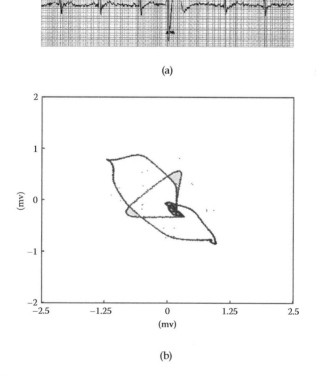

FIGURE 8.11
ECG summary report. (Redrawn from Fan, X., He, C., Cai, Y. and Li, Y., in *IEEE CloudCom 2012*, Taipei, December 3–6, 2012, 705–710.)

(a)

(b)

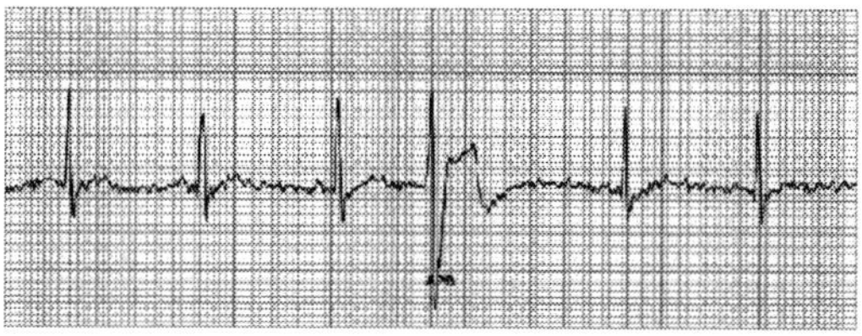

FIGURE 8.12
(a) Abnormal ECG. (b) Chaos analysis.

8.3.2.1 Calculation of PPG Data

A PPG is an optically obtained plethysmogram, which represents a volumetric measurement of an organ. It is often obtained using a pulse oximeter that illuminates the skin and measures changes in light absorption [21–23]. It is used to monitor conditions related to breathing, hypovolemia, and other circulatory situations. The HCloud system provides the heart function indices, which evaluate the heart's blood pumping capability, and peripheral vascular function, which assesses for HBP and arteriosclerosis. Two assessments of PPG are delivered to users: heart function and peripheral vascular function. Meanwhile, the system provides a detailed PPG index on heart function for the physician, including average pulse rate (PR), cardiac output (CO), stroke volume (SV), blood oxygen saturation, and cardiac index (CI). To make users and the physician understand the heart function parameters intuitively, this system provides each index specification as well as the index of heart function histogram, as shown in Figure 8.13a. It is known that the index of peripheral vascular function can reflect the health status of the peripheral vascular system, which can help a physician assess serious degrees of HBP and arteriosclerosis. On the other hand, this health care cloud platform provides users the waveform characteristic (K), blood viscosity (V), peripheral resistance (TPR), sclerosis index (SI), degree of vascular conformity (AC), and pulse wave transit time (PWTT). At the same time, it plots the peripheral vascular histogram and gives specifications, as shown in Figure 8.13b. From those indices, a PPG diagnostic report can be generated by the system for the end user.

8.3.2.2 Presentation of HBP Signals

High blood pressure or hypertension is a chronic medical condition in which the blood pressure in the arteries is elevated, which causes the heart to work harder than normal to circulate blood through the blood vessels. Blood pressure involves two measurements: systolic and diastolic. These measurements depend on whether the heart muscle is contracting (systole) or relaxing in the interval between beats (diastole). Normal blood pressure at rest is within the range of 100–140 mm Hg systolic (top reading) and 60–90 mm Hg diastolic (bottom reading) [24]. HBP is said to be present if the blood pressure reading is persistently at or above 140/90 mm Hg, which can cause problems with the metabolism of fat and sugar, as well as changes of the heart, brain, kidneys, and retina. An HBP report is provided in HCloud to provide detailed data on the systolic and diastolic blood pressure and inform users whether they suffer from HBP. Over the long term, it can reflect that the degree of control of HBP may be associated with the health status of the user and prompt the user to change his or her lifestyle. Due to the absence of obvious clinical symptoms in some patients with HBP, it is known as the "invisible killer" [25]. Fortunately, HCloud can draw a curve of HBP for a long-term trend of threatening status.

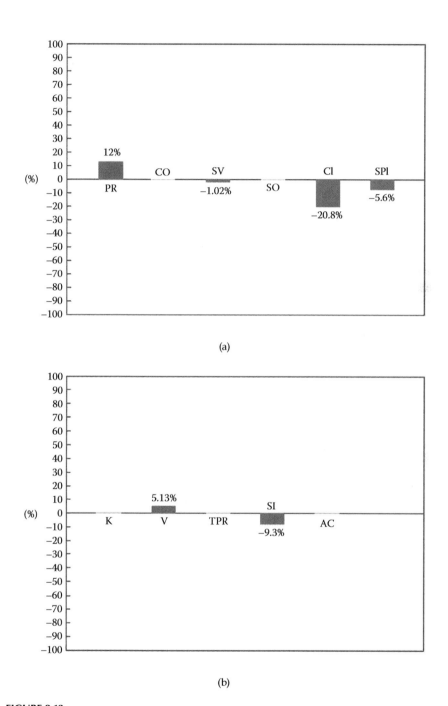

FIGURE 8.13
PPG data indicies report: (a) heart function index; (b) peripheral vascular index. (Redrawn from Fan, X., He, C., Cai, Y. and Li, Y., in *IEEE CloudCom 2012*, Taipei, December 3–6, 2012, 705–710.)

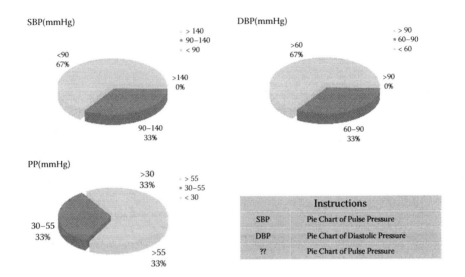

FIGURE 8.14
Pie chart for HBP index.

It is extremely important for people to be aware of HBP so there can be early prevention and timely treatment. In our system, the HBP report is provided to provide detailed data on systolic and diastolic blood pressure and inform users whether their blood pressure is high or not. Figure 8.14 shows the distribution of the range of HBP in one period, which helps users monitor their HBP status. At the same time, it can help a doctor know the general health state of a patient. A trend chart of HBP (Figure 8.15) shows which phase blood pressure pills will affect.

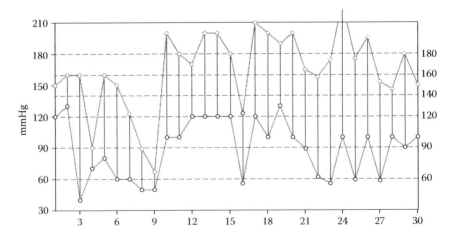

FIG 8.15
Trend chart for HBP.

8.4 Performance Testing and Evaluations

8.4.1 Case Design and Simulation

To evaluate the performance of HCloud, we designed two cases to simulate the most-used scenarios: uploading information and browsing results. The testing environment included 10 personal computers (PCs) as clients with CentOS 6.0 OS (operating system), Core 2.8-GHz CPU, and 2 G memory for each PC. Five virtual machines were also established to enhance the utilization of the physical host in five physical machines. Each virtual machine had the same configuration (Core 2.8-GHz CPU, 1 G RAM). All of the testing clients, using the Tsung testing tool [26], would simulate high concurrent access in the real world to perform stress and load testing. In actual application, people can press buttons on a remote to process the uploading procedure via an upload interface on a TV as shown in Figure 8.16.

The first test verified the capability of the platform for uploading data. One hundred concurrent users were generated by simulation and uploaded data to the servers continuously for 10 minutes. Among these 100 users, 60 uploaded ECG data, 30 uploaded PPG data, and the last 10 uploaded blood pressure values. Each of the ECG and PPG record's length was 2 minutes. As the test results show in Figure 8.17 and Table 8.3, we can summarize that the test produced a total of 158,780 requests and nearly 30,000 concurrent connections were maintained at the end of the test; each single server traffic spike was 5 Mb/s. Only one response failed during the test. The next test verified the concurrent capability of the web servers. Three hundred simulated users were generated to browse the web page at the same time, with each visit

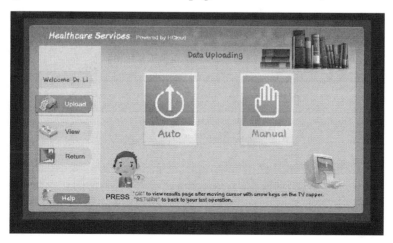

FIGURE 8.16
Presentation of a user's upload interface on a TV. (Redrawn from Fan, X., He, C., Cai, Y. and Li, Y., in *IEEE CloudCom 2012*, Taipei, December 3–6, 2012, 705–710.)

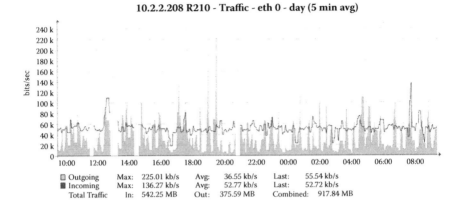

FIGURE 8.17
Inbound and outbound traffic of the network for uploading. (Redrawn from Fan, X., He, C., Cai, Y. and Li, Y., in *IEEE CloudCom 2012*, Taipei, December 3–6, 2012, 705–710.)

TABLE 8.3

Uploading Activity Statistic for HTTP Request

HTTP Status Code	200 (Response Is OK)	302 (Redirect Is OK)	500 (Internal Server Error)
Highest rate (bytes/second)	581.4	1.8	0.1
Total number	158,780	162	1
Finish_users_count (the total number of concurrent connections): 28,170			

lasting 30 minutes. Among these 300 users, 100 browsed the left navigate list, 100 visited each icon, and 100 browsed the main content links. In the testing results shown in Figure 8.18 and Table 8.4, we summarize that the testing generated 550,636 requests, the number of concurrent connections exceeded 20,000, and each single server traffic spike was approximately 30 Mb/s in this situation.

8.4.2 Results Evaluation

The connection status and throughput in the simulation provide a view of the system performance. Further details and evaluation are depicted in this segment as follows: Table 8.5 shows the details of requests that happened for uploading and browsing. A new HTTP *request* generated within a given interval of 0.02 seconds represents that a new user connection happened, and a *session* will be created according to the probability presented in the testing configuration file. The mean response time and count (for page, request, etc.) for the entire test are computed, generating 163,011 and 573,683 concurrent requests for the two activities, respectively, as shown in *Mean*

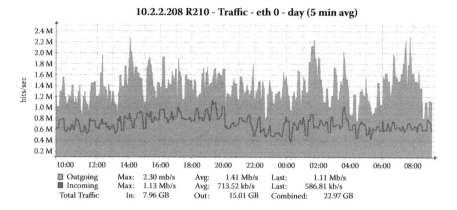

FIGURE 8.18

Inbound and outbound traffic of the network for browsing. (Redrawn from Fan, X., He, C., Cai, Y. and Li, Y., in *IEEE CloudCom 2012*, Taipei, December 3–6, 2012, 705–710.)

TABLE 8.4

Browsing Activity Statistic for HTTP Request

HTTP Status Code	200 (Response Is OK)	302 (Found)
Highest rate (bytes/second)	2,568	99.9
Total number	550,636	22,810
Finish_users_count: 23,233		

TABLE 8.5

Overall Performance Statistics for Activities

Name	connect	page	request	session
Uploading Activity				
Highest rate (bytes/second)	118.9	243.6	604.7	103.6
Mean time (seconds)	0.12	0.40	0.16	1.22
Count	32,225	64,771	163,011	28,170
Browsing Activity				
Highest rate (bytes/second)	121.9	896.4	2611.1	98.4
Mean time (seconds)	0.3	0.25	0.08	14.98
Count	23,633	206,381	573,683	23,233

Source: Redrawn from He, C., Fan, X., and Li, Y., *IEEE Transactions on Biomedical Engineering* 60, no. 1 (January 2013): 230–234.

Note: *Connect*, duration of the connection establishment; *page*, response time for each set of requests; *request*, response time for each request; *session*, duration of a user's session.

TABLE 8.6

Network Traffic and Performance Statistics for Activities

Receive/Sent	Activity	Bandwidth (KB/seconds)	Total (MB)
size_rcv (size of responses)	Uploading	835.84	224.58
	Browsing	26,277.12	6,113.28
size_sent (size of requests)	Uploading	2,999.04	803.37
	Browsing	522.24	108.54

Source: Redrawn from He, C., Fan, X., and Li, Y., *IEEE Transactions on Biomedical Engineering* 60, no. 1 (January 2013): 230–234.

TABLE 8.7

Statistics for Reading Performance of Amazon S3

Page Size (KB)	10	100	1,000
Response time (seconds)	0.14	0.45	3.87
Bandwidth (KB/seconds)	71.4	222.2	258.4

Source: Brantner, M., Florescu, D., Graf, D., Kossmann, D., and Kraska, T., in *Proceedings of the 2008 ACM SIGMOD International Conference on Management of Data*, Vancouver, B.C., Canada, June 9–12, 2008, 251–263.

Time and *Count* rows. A response time less than 1 second is tolerable for the user's experience. The actual number of concurrent users was nearly 30,000 during the experiment, as the *Finish_users_count* column of Table 8.3 and Table 8.4 shows. So, it is acceptable to actual applications.

By comparison, Table 8.7 shows the reading performance of Amazon's S3 with different page sizes (writing data to S3 takes about three times as long as reading data) [27]. Ignoring the slight differences of hardware, the usability of our system is satisfied in terms of *Bandwidth*, which is at least double that in S3 (compare the last row of Table 8.6 with that of Table 8.7). *Response Time* is also better, obviously. Overall, it demonstrates that this document-oriented cloud storage architecture is more appropriate for the health care services environment with a large number of trivial files rather than non-document-oriented ones such as S3. Furthermore, due to platform capabilities of linear extendibility, we simply increase the number of servers if there is the potential for more users' requests in the future.

8.4.3 Comparison of Computing Overhead

From Section 8.4.2, we can find that the response time is acceptable to our applications from the point of view of accessing HTTP. Actually, the main tasks of HCloud are physiological data processing and computing, as discussed previously, which can affect the performance of the whole system. We also take the ECG raw data process as an example. Using a typical parallel

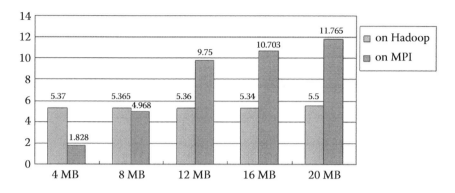

FIGURE 8.19
CPU time spent on Hadoop and MPI.

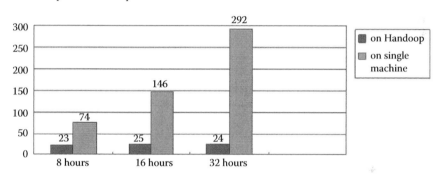

FIGURE 8.20
Running time on Hadoop and a single machine.

framework message passing interface (MPI) program model, CPU time spent filtering baseline wander is compared with that of Hadoop. In comparison, a single machine computing overhead is depicted in Figure 8.19.

On Hadoop, with the increase of data, the running time is essentially the same. On a single machine, the time changes linearly with the amount of data. For example, with 32 hours of data, the performance is improved 11 times. For three nodes, the performance is improved by 82%. So, this platform achieves higher performance than ordinary solutions, as shown in Figure 8.20.

8.5 Conclusion

The HCloud system incorporates multiple advanced technologies, such as a precise and convenient data acquisition solution and a high-efficiency data storage and analysis method, for monitoring the health status of users at any time and in any place. Through the system, anyone can have knowledge regarding personal health information and even be told the risk of some chronic diseases in the future. With our system, some acute attacks can be discovered in time, and chronic diseases such as hypertension can be prevented before their onset. This chapter proposed a private cloud platform architecture associated with technologies such as MQ, load balance, session cache, and cloud storage. This platform can integrate semistructured, unstructured, and heterogeneous physiological signal data well and can support huge data storage and heterogeneous data processing for various health care applications, such as automated ECG analysis, PPG analysis, and HBP analysis. It is also a low-cost solution that can reduce module coupling by adopting component technology. Moreover, the proposed system can provide an early warning mechanism for people with chronic diseases and help physicians obtain patients' health information. The Map-Reduce paradigm has the features of code simplicity, data locality, and automatic parallelization compared with other distributed parallel systems. More important, integrated with the HCloud is improved efficiency of physiological data processing and achievement of linear speed-up. Based on the performance evaluation and feedback from user experiences, HCloud can cope with the issues of high concurrent requests in ubiquitous health care service and dispose of the analysis of massive physiological signal tasks quickly, as well as having robust, instant, and efficient features that can meet user demands for preventive health care.

References

1. Ding, H., He, J., and Wang, W., The sub-health evaluation based on the modern diagnostic technique of traditional Chinese medicine, in the *Proceedings of the First International Workshop on Education, Technology and Computer Science, ETCS '09*, Editors Hu, Z. and Liu, Q., Wuhan, China: Printing House, March 7–8, 2009, 269–273.
2. World Health Organization (WHO), World health statistic 2011. http://www.who.int/whosis/whostat/2011/en/ (accessed April 5, 2013).
3. Robison, S., The next wave: everything as a service. 2007. http://www.hp.comlhpinfo/execteam/articles/robison/08eaas.html (accessed April 5, 2013).
4. Linthicum, D. S., *Cloud Computing and SOA Convergence in Your Enterprise: A Step-by-Step Guide*. Boston: Addison-Wesley, 2009.
5. Jericho Cloud Cube. http://www.jerichoforum.org/cloud_cube_model_v1.0.pdf (accessed November 2, 2012).
6. DMTF, *Interoperable Clouds*. White paper. November 2009. http://www.dmtf.org/sites/default/files/standards/ (accessed April 5, 2013).
7. Samba, A., Logical data models for cloud computing architectures, *IT Professional* 14, no. 1 (2012): 19–26.

8. Fan, X., He, C., Cai, Y. and Li, Y., HCloud: a novel application-oriented cloud platform for preventive healthcare, *IEEE CloudCom 2012*, Taipei, December 3–6, 2012, 705–710.

9. He, C., Fan, X., and Li, Y., Toward ubiquitous healthcare services with a novel efficient cloud platform, *IEEE Transactions on Biomedical Engineering* 60, no. 1 (January 2013): 230–234.

10. Zhang, R., and Liu, L., Security models and requirements for healthcare application clouds, *The 2010 IEEE 3rd International Conference on Cloud Computing (CLOUD)*, July 5–10, 2010, Miami, FL, USA, 268–275.

11. Narayanan, H., and Gunes, M., Ensuring access control in cloud provisioned healthcare systems, *IEEE Consumer Communications and Networking Conference (CCNC)*, Jan. 8–11, 2011, Las Vegas, NV, USA, 247–251.

12. Chang, H. H., Chow, P. B., Ramakrishnan, S., et al., An ecosystem approach for healthcare services cloud, *IEEE International Conference on e-Business Engineering*, Oct. 21–23, 2009, Macau, China, 608–612.

13. GlusterFS. http://www.gluster.org/about (accessed April 5, 2013).

14. Stonebraker, M., SQL databases v. NoSQL databases, *Communications of the ACM* 53, no. 4 (April 2010): 10–11.

15. Scully, C. G., Lee, J., Meyer, J., Gorbach, A. M., Granquist-Fraser, D., Mendelson, Y., and Chon, K. H., Physiological parameter monitoring from optical recordings with a mobile phone, *IEEE Transactions on Biomedical Engineering*, 59, no. 2 (February 2010): 303–306.

16. MQ API Guide. http://www.rabbitmq.com/api-guide.html (accessed April 5, 2013).

17. Ronald C Taylor, An overview of the Hadoop/MapReduce/Hbase framework and its current applications in bioinformatics, editor, Ronald C. Taylor, in *Proceedings of the 11th Annual Bioinformatics Open Source Conference*, July 9–10, 2010, Boston, MA, USA.

18. Aswini Kumar, ECG-simplified. http://www.lifehugger.com/doc/120/ecg-100-steps (accessed April 5, 2013).

19. Fen Miao, Xiuli Miao, Weihua Shangguan, and Ye Li, MobiHealthcare System: Body Sensor Network Based M-Health System for Healthcare Application, *E-Health Telecommunication Systems and Networks*, 1(1), 2012, 12–18.

20. Huang, X., Li, T., Dai, H., and Li, Y., The parallel processing of ECG signal based on Hadoop framework, in *Proceedings of the 9th International School and Symposium on Medical Devices and Biosensors and the 8th International School and Symposium on Biomedical and Health Engineering (MDBS-BHE 2013)*, Hong Kong, June 27–30, 2013.

21. Luo, Z. H., Zhang, S., and Yang, Y. M., *Engineering Analysis for Pulse Wave and Its Application in Clinical Practice*. Beijing: Science Press, 2006.

22. Shelley, K., and Shelley, S., Pulse oximeter waveform: photoelectric plethysmography, in *Clinical Monitoring*, C. Lake, R. Hines, and C. Blitt, eds. Philadelphia: Saunders, 2001, 420–428.

23. Shamir, M., Eidelman, L. A., Floman, Y., Kaplan, L., and Pi-zov, R., Pulse oximetry plethysmographic waveform during changes in blood volume, *British Journal of Anaesthesiology*, 82 (1999): 178–181.

24. Carretero, O. A., and Oparil, S., Essential hypertension. Part I: definition and etiology, *Circulation* 101, no. 3 (2000): 329–335.

25. Chobanian, A. V., Bakris, G. L., Black, H. R., et al., Seventh report of the Joint National Committee on Prevention, Detection, Evaluation, and Treatment of High Blood Pressure, *Hypertension* 42, no. 6 (2003): 1206–1252.
26. Tsung user's menu. http://tsung.erlang-projects.org/ (accessed April 5, 2013).
27. Brantner, M., Florescu, D., Graf, D., Kossmann, D., and Kraska, T., Building a database on S3, in *Proceedings of the 2008 ACM SIGMOD International Conference on Management of Data*, Vancouver, Canada, June 9–12, 2008, ACM, 251–263.

9

RPig: Concise Programming Framework by
Integrating R with Pig for Big Data Analytics

MingXue Wang and Sidath B. Handurukande

CONTENTS

Summary

In many domains, such as telecommunications, various scenarios necessitate the processing of large amounts of data using statistical and machine learning algorithms for deep analytics. A noticeable effort has been made to move the data management systems into MapReduce parallel processing environments, such as Hadoop, and Pig. Nevertheless, these systems lack the necessary statistical and machine learning algorithms and therefore can only be used for simple data analysis. Frameworks such as Mahout, on top of Hadoop, support machine learning, but their implementations are at the early stage. For example, Mahout does not provide support vector machine (SVM) algorithms, and it is difficult to use. On the other hand, traditional statistical software tools, such as R, containing comprehensive statistical algorithms for advanced analysis, are widely used. But, such software can only run on a single computer; therefore, it is not scalable for big data. In this chapter, we present RPig, an integrated framework with R and Pig for scalable machine learning and advanced statistical functionalities, which makes it feasible to use high-level languages to develop analytic jobs easily in concise programming. Using application scenarios from the telecommunications domain, we show the use of RPig. With comparable evaluation results, we demonstrate advantages of RPig, such as less development effort compared with related work.

9.1 Introduction

With the explosive growth in the use of information communication technology (ICT), applications that involve deep analytics need to be shifted to scalable solutions for big data. Our work is motivated by the big data analytic capabilities of network management systems, such as network traffic analysis, in the telecommunications (telecom) domain. More specifically, the work is an extension of Apache Pig/Hadoop frameworks, which are commonly used to build cost-effective big data systems in industry. The design, the developed software implementation, and the solution we describe here are general and applicable to other domains.

To build a scalable system, one approach is to use distributed parallel computing models, such as MapReduce [1], that allow adding more (computer) nodes into the system to scale horizontally. MapReduce has been recently applied to many data management systems (DMSs), such as Hadoop and Pig. These systems target the storage and querying of data for top-layer applications. However, they lack the necessary statistical and machine learning algorithms and therefore can only be used for simple data analysis. For advanced or deep analysis, Mahout [2] contains a limited number of machine learning algorithms implemented in the MapReduce model. Because of the large number and complexity of machine learning and statistical algorithms, the redesign and redevelopment of these algorithms in the MapReduce model are difficult tasks. Various algorithms are still missing in Mahout in comparison with matured statistical and machine learning frameworks. For example, support vector machines (SVMs), one commonly used algorithm, is still under development in Mahout. On the other hand, traditional statistical software, such as R, has a rich and extensive set of machine learning and statistical processing functionalities for advanced analysis, but it is not distributed and not scalable on its own. In general, it only runs within a single computer and requires all data to be loaded into memory for processing. Some solutions have been proposed to scale out this traditional statistical software, such as RHadoop [3], but limitations still exist. For example, some require writing key-value paired map and reduce functions, leading to difficulties in use and longer development time. More details of related work are described in Section 9.6. Our approach addresses the problem by integrating traditional and matured statistical software (R) with a scalable DMS (Pig) to scale out deep analytics.

In this chapter, we present RPig, an integrated framework with R and Pig for scalable machine learning and advanced statistical functionalities, which makes it feasible to use high-level languages to develop analytic jobs easily in concise programming. RPig takes advantage of both the deep statistical analysis capability of R and parallel data-processing capability of Pig. Both data storage and processing for deep data analysis are distributed and scalable. The framework has the following main advantages:

- The statistical and machine learning functions of R can be easily wrapped and directly used with Pig statements. This allows developing advanced parallel analytic jobs with two high-level languages R and Pig (Latin) without needing to learn new languages or application programming interfaces (APIs) or rewrite complex statistical algorithms. The development effort can be significantly reduced for the user.

- The framework is able to parallelize both R and Pig executions automatically at the execution stage. The necessary low-level operations, such as data conversion and fault handling, are handled by the framework itself. The framework offers automatic parallel execution for advanced data analysis.

In the rest of the chapter, we describe two scenarios that we encounter in Section 9.2 that neither R nor Pig can handle independently. Section 9.3

describes the foundation frameworks: R, Hadoop, and Pig. The overall RPig framework and its components are explained in Section 9.4. Experiments and results are in Section 9.5. Finally, we talk about related work and give our conclusion (Sections 9.6 and 9.7, respectively).

9.2 Motivating Scenarios

To demonstrate the need and usefulness of our RPig framework, we describe two example use cases in the context of network management systems where scalable statistical processing is necessary.

9.2.1 Intensive Scenario with Both Input/Output and Central Processing Unit with Exponential Moving Average

In this first use case, a vast amount of events are collected from a given mobile network and stored as event log files. An event is a report about a particular service client (e.g., Viber voice over Internet protocol [VoIP] service client) and contains information such as

```
ID|period_start|period_end|IMSI|IMEISV|RAT|...
  |packets_downlink|packets_uplink|...
```

The exponential moving average (EMA) is a simple forecasting algorithm based on historical sample data. Using the EMA, an analytic feature of a network management system can forecast the amount of traffic of selected service clients in the next time window when a request is sent. Because of the vast number of events, it is impossible for R to load all data into memory for a simple EMA calculation. However, Pig does not have the EMA function, which R has.

This problem can be addressed by RPig, which allows log files to be efficiently loaded, preprocessed (filtering, aggregating, etc.) by Pig in parallel, and then directly passes the data to R for a final EMA calculation. In this case, it is both an input/output (I/O) and central processing unit (CPU) intensive scenario as it requires loading and preprocessing massive log files from hard disks.

9.2.2 A CPU-Intensive Scenario with SVM

The SVM machine learning algorithms can be used for advanced classification and regression analysis. Unknown data can be predicted by an SVM model, which is built from training data in the training phase.

An increasing amount of phone calls are made by various VoIP clients, such as Viber and Skype. One approach for monitoring the service quality of VoIP is using network-level key performance indicators (N-KPIs) at the Internet protocol (IP) layer, such as packet loss or jitter, to predict the mean opinion score (MOS), which is a standard speech quality measurement parameter [4]. An SVM-based regression algorithm is used in this case, but it is a complex algorithm, usually involving long computation times on a relatively small amount of data in the training phase. RPig enables us to define and execute the SVM algorithms in the MapReduce model for both SVM training and prediction phases without writing any key-value pair MapReduce functions. As a result, the performance becomes scalable to cluster size, and development effort is reduced.

This use case deals with a complex machine learning algorithm, which is CPU intensive rather than I/O intensive. R's in-memory computation takes most of the overall computation time with a few data in an analysis job. RPig supports parallelism for various requirements in different scenarios.

9.3 Background

Big data [5] are data in volumes so large and complex that they become difficult to process using on-hand database management tools or traditional data-processing applications. Since Google published its MapReduce technology and Apache started the Hadoop project in 2004 and 2005, MapReduce and Hadoop have become a generic and foundational approach for developing scalable, cost-effective, flexible, fault-tolerant big data systems [6]. Many frameworks, such as Pig and Hive, have been developed based on Hadoop, adding features on it. As Hadoop systems are more widely adopted in industry, the requirements of the real-world problems are driving the Hadoop ecosystem to become even richer. For example, Oozie and Azkaban provide workflow and scheduling management. Impala and Shark aim at low-latency real-time queries. Our work, RPig, is one of many frameworks, such as Mahout and DataFu [7], targeting deep analytics. In the following sections, we briefly describe the frameworks on which the RPig is based.

9.3.1 R and R Packages

R is a programming language and software environment widely used for statistical computing and deep data analysis, such as classification, and regression. R is extensible through R packages. There are thousands of R packages that implement massive specialized machine learning and statistical algorithms.

R's data model contains simple data structure types, such as *scalars, vectors,* and *lists,* and special compound data structure types: *Factors* are used to describe items that can have a finite number of values; *data frames* are *matrices* and may contain different data types (numeric, factor, etc.). All data structures of R are R objects, which also include other statistical specific models or functions and so on.

The following code snippet shows a simple example of EMA calculation using R. *TTR* is an R package implementing various moving average calculations. The *temp* is a series for EMA calculation with 20 periods to average over.

```
Library(TTR); results <- EMA(temp, 20)
```

9.3.2 Hadoop and MapReduce

Hadoop offers the Hadoop Distributed File System (HDFS) to manage data storage and a distributed parallel programming framework based on MapReduce [5] for data processing. Computations are defined in *Map* and *Reduce* functions, which have key-value pairs for input. A map function takes one pair of data, which can be processed in parallel Map(k1,v1)→list(k2,v2). A reduce function aggregates related results of map functions (k2, list(v2))→list(v3). Programs need to be written as map and reduce programs to enable parallel computing through Hadoop MapReduce Java APIs.

9.3.3 Pig and Pig Latin

Pig is built on top of Hadoop and gives a high-level data flow language called Pig (Latin) [8] for expressing data queries and processing. It is similar to SQL of a relational database management system (RDBMS), but it is procedural style and gives more control and optimization over the flow of the data. Pig scripts are compiled into sequences of MapReduce jobs by Pig, and they are executed in the Hadoop MapReduce environment.

The Pig data model contains scalar types that have a single atomic value (integer, long, etc.), and three complex types that can contain other types: Tuple is a data record consisting of a sequence of "fields," which can be any data type; Bag is a set of tuples, similar to a "table"; Map is a map of a string key to a value, which can be any data type.

Pig provides a set of operators for data processing. For example: LOAD and STORE can be used for reading and writing data from HDFS. Processing every tuple of a data set can use the FOREACH operator. Many operators are similar to SQL, such as JOIN, GROUP BY, and UNION for standard data operations. As with many SQL implementations, Pig supports user-defined functions (UDFs), which allows performing tasks written in low-level language (Java or Python) to extend Pig. The following Pig script shows how to aggregate traffic consumption (both up/downlink) on selected VoIP clients (e.g., Skype, Viber) in a time window on events described in Section 9.2.1.

```
Events = LOAD '$load_par' USING PigStorage('|') AS
  (ID, period_s:LONG,...,);
Events = FILTER Events BY (client = = 'Viber' OR...);
Traffics = FOREACH (GROUP Events BY (period_s, period_e, client))
  GENERATE FLATTEN (group), (SUM(Events.downlink)+SUM(Events.
  uplink)) AS links:DOUBLE;
```

9.4 The Framework

An initial version of the RPig framework [9] was implemented as a proof-of-concept prototype. The framework provides the RPig script for users to write analytic jobs. The RPig script inherits Pig script syntaxes as the language skeleton but allows defining inline R scripts as R functions. An R function element will be interpreted as an input payload of a predefined Pig extended function or Pig UDF, which handles the payload at the execution stage. This design gives us a quick implementation by only using the Pig UDF APIs without going through the Pig source code. However, it is not an optimal approach for integrating Pig and R. RPig script has its own constructs, and it needs to generate additional Pig supporting statements in execution. The initial version also has the large performance overhead of the data exchange between R and Pig.

To improve the performance of RPig and to integrate R and Pig in an optimal way, we completely redesigned and rewrote the source code to overcome the aforementioned disadvantages of the initial version. By doing so, we have brought the research prototype to an early production stage. Some of the main advantages of the current version over the initial proof-of-concept version are the following:

- There is seamless integration with Apache Pig by having a built-in R script extension similar to other Pig script extensions, such as Python and JavaScript.

- Only standard R and Pig language syntaxes are used without any new language constructs. It allows the use of any existing R and Pig script integrated development environment (IDEs).

- There is support for two types of R engines. R UDFs can be executed on the Java virtual machine (JVM) or a stand-alone R engine.

- Much faster performance is provided. Optimized data conversion and verbosity XML (extensible markup language) messages are not involved as the intermediate data format.

In the following sections, we describe the current version of the framework in detail.

9.4.1 The R Script Engine Extension

To integrate R and Pig and take advantage of both, the R language is expected to be supported to define Pig UDFs for specifying custom processing in Pig data flows. Pig already supports a number of languages, such as Python and JavaScript for UDFs. They are implemented as different script engine extensions in Pig. That is, an R script engine extension (RScriptEngine) is required for our case. It wraps the R engine in the back end, which can interpret R scripts at runtime (Figure 9.1). The user defines R functions as UDFs in an R script and makes Pig aware of the R script by using the Pig REGISTER statement in a data flow (step 1 of Figure 9.1). An RScriptEngine will be initialized, and it will register the defined R functions. The RScriptEngine will be shipped within Pig-generated MapReduce programs to all Hadoop task nodes during execution (step 2 of Figure 9.1). RScriptEngine can execute the registered R functions in the back-end R engine by providing a bridge function for interactions between Pig and R (step 3 of Figure 9.1).

The back-end script engine is usually selected from the Java implementations of the script language. For example, Jython and Rhino are used for Python and JavaScript back-end engines, respectively. This enables running the script languages on the JVM where Hadoop and Pig are running. Hence, no additional back-end script engine is required to be installed on every host along with the

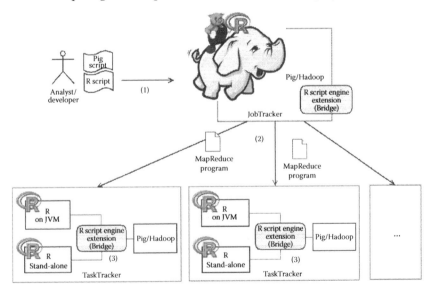

FIGURE 9.1
The framework overview.

JVM. However, this is not a case for the R language since there is no mature JVM-based R interpreter. Some preliminary implementations are available, such as Renjin [10], but it is incompatible with most R packages except for some basic libraries. As a result, two types of back-end R engine are supported in our implementation (JVM based and stand-alone R). JVM-based R can be used for some basic statistical functions (e.g., standard deviation) without requiring the installation of a script engine on all nodes. Stand-alone R is able to use any R functions and packages. However, the R must be preinstalled on all hosts.

9.4.2 Data-Type Conversions

It is necessary for data to be passed back and forth between Pig and R during the R function executions. Since the two languages Pig and R have very different data models, the data must go through a conversion process, which is one of the main responsibilities of the R bridge function. The data-type conversion is done automatically based on the set of predefined rules discussed next.

9.4.2.1 *From Pig to R*

- Simple data type
  - `int:` integer; `long/float/double:` double; `chararray/datetime:` character `bytearray:` raw; `boolean:` logical (e.g., null: NULL); `datatime:` POSIXlt,POSIXt;
- Complex data type
  - `tuple:` list, e.g. (19,2): list(19,2); `dataBag:` nested list, e.g. {(19,2), (18,1)}: list(list(19,2), list(18,1)); `map:` named list, e.g., [apache#pig]: list(key = "apache", value = "pig")
- Anything else raises an exception

Any nested data objects in Pig, such as nested tuples, will be converted to nested lists in R. Due to the different purposes of the two languages, there is no exact semantic match between all data types in their data models. For example, the `map[key#value]` type of Pig is hardly used in statistical computing, so we convert it to a named `list(key = key, value = value)`, which is an ordered collection in R. Users can still convert the converted R object to other R data types via R operations (inside R functions) if necessary. For example, it would be possible to convert a nested list to a data frame or a matrix.

9.4.2.2 *From R to Pig*

When the data must be sent back to Pig after R execution, a user-defined output schema of the R function is needed. This allows the user to specify what they expected from the R output and remove ambiguity during the conversion. For example, the *logical* value in R could be True, False, or NA (Not Available), but the Pig *Boolean* type can only be either True or False. By using the output schema, the *logical* value can be converted to a Boolean value. Alternatively, the user may specify an *int* or *chararray* value and no semantic information is lost. The following rules are used for type conversion from R to Pig:

- Simple data type
    - (schema: int) `numeric/integer/logical/factor:` `int` (T:1; F:0; NA:128); (schema: float/double) `numeric/double:` `double`; (schema: chararray) `character/logical/factor:` `chararray`; (schema: bytearray) `raw:` `bytearray`; (schema: boolean) `logical:` `boolean` (T: T, F/NA: F); `NULL:` `null`; (schema: datetime) `POSIXlt/POSIXt:` `datetime`
- Complex data type
    - (schema: tuple) `numeric array/character array/logical array/factors/list:` `tuple`, e.g. `structure(c(1L, 2L, 1L)`, .Label = c("a", "b"), class = "factor"): (a, b, a); (schema: bag) `nested list:` `bag`; (schema: map) `list:` `map`
- Anything else raises an exception.

9.4.3 Execution and Monitor

At the parallel execution stage, the defined R functions or UDFs are transformed into map functions that are automatically generated by taking advantage of Pig. They are executed in parallel in different Hadoop task nodes. Each R or map task will take a piece of split data and execute independently on an R engine on one task node. If the Hadoop cluster is configured with more than one map task capacity per node, each map or R function will have an isolated session. When a task is completed and a result is returned, the data stored in the R session will be cleared, and the process will be killed by the RPig framework. As a consequence, no R session will be kept alive after the R execution is complete, and all data that need to be saved or persisted from R must be saved in HDFS through Pig operations. This design was chosen because an R session only exists in a single task node, which is replaceable by any other task node in a Hadoop cluster at any time. The R session cannot be retrieved by other nodes at a later time. Pig stores data, including temporary data generated between MapReduce jobs during processing, in HDFS to guarantee that data can be retrieved later from every node of the cluster. The results of all R functions will be collected through Reduce tasks for continuous processing. Users do not need to develop key-value pair map and reduce functions within RPig. They only need to assign the number of map and reduce tasks in parallel execution through Hadoop and Pig configuration.

With regard to fault tolerance, the fault handling happens in two different layers, the node layer and the R engine layer. The underlying Hadoop framework provides failure handling on nodes of the cluster. If a node fails during the execution of an RPig job, Hadoop will restart the task of the failed node in an alternate node. Within a node task, the RPig framework allows the user to define the fault policy to handle errors from an R engine execution on an R function. For example, by default `func _ name.fault.ignore←T`. This policy ignores any exceptions and continues. Also, `func _ name.fault.retry←1`. This allows at most one retry when an exception occurs. If an R execution fails during the map task, then a remedy action defined in a failure policy of the named R function will be applied, and the failure event will be logged by the RPig framework. The user still can use R's `tryCatch()` function within an R function to define the fault-handling mechanisms within the R session, but the fault policy of RPig allows the user to restart the R function in a brand new R session.

R functions may run exceedingly slowly on occasion, and the user would expect a way to monitor the UDF execution time and terminate its execution if it runs too long. RPig offers the facility for monitoring long-running R functions. For example, `func _ name.monitoredUDF.duration←10` will terminate the named R function if it runs for more than 10 seconds and return the default value of null.

9.4.4 Implementation

There are several libraries used for the RPig implementation. Renjin [10] is used for the JVM-based R engine. Since the stand-alone R is implemented in C and Fortan and Pig is written in Java, Rsession [11] is adopted as the Java interface of R to use the Pig APIs. Pig offers Java annotation-based implementation for a monitored Java UDF. To build the same function for R UDFs, we need to create a new Java class with annotations for each R function at runtime. The Javassist [12] is used for defining a new class at runtime and to modify a class file when the JVM loads it.

9.5 Use Case and Experiment

In this section, we describe the usage of RPig with the examples we discussed in Section 9.2. To provide valuable comparative experimental results, we also describe and experiment with one alternative framework or implementation for each use case. Although the use cases here are from the telecom domain, the design and the solution we describe are general and applicable to other domains.

Our experiments are conducted in Amazon Elastic MapReduce (EMR), for which we have all nodes with the same configuration (m1.medium instance,

Hadoop 1.0.3, R 2.14.1). One node from the Master Instance Group has the extended Pig 0.11 with the RPig feature deployed to generate MapReduce programs. The rest of the nodes are from the Core Instance Group, providing both data storage and MapReduce task execution services. As R requires data to be loaded in memory, each node is configured to have a maximum capacity of one map task and one reduce task, so an R session could take the maximum memory available in a single node. We also assign a larger heap-size limit to the child JVMs of map tasks as these are where R statistical functions are executed. The reduce task is allocated a lower value.

9.5.1 Summary Statistics with Quantiles

Before going to complex examples that use different R packages, we would like to show a simple quantiles statistic task to give a "hello world" example in the first case. Quantiles are used to summarize a set of observations by giving the boundary values between the divided distributions. For example, a large number of values for a network parameter observed over time can be summarized in a few numbers, or quantiles, for reporting or comparing with thresholds.

9.5.1.1 Design and Implementation

DataFu [7] is a collection of useful Pig add-ons (UDFs) developed by LinkedIn for data mining and statistics, and it is used for the comparison study in this use case. DataFu is used in many off-line workflows for data-derived products like "People You May Know" and "Skills" at LinkedIn. The following shows the main lines of implementation using DataFu.[*]

```
DEFINE Quantile datafu.pig.stats.Quantile('0.5','0.75','1.0');
— Computing the quantiles for each network nodes
Quantiles = FOREACH B {sorted = ORDER values BY val; GENERATE
  id, Quantile(sorted); };
```

DataFu uses the DEFINE statement to specify a Quantile UDF function with string parameters for the function constructor. ('0.5','0.75','1.0') yields the 50th and 75th percentiles and the max. The function takes a sorted bag as the input.

The following shows the RPig version for the same computational task.

```
REGISTER 'RFuncs.r' using rsession as RFuncs; — or using
  'renjin' for JVM R
%DECLARE q_probs '0.5, 0.75, 1';%declare q_type '2';
Quantiles = FOREACH A GENERATE id, RFuncs.Quantile(values,
  '$q_probs', '$q_type');
```

---

[*] Detailed explanation of the DataFu quantile example is available at http://engineering.linkedin. com/open-source/introducing-datafu-open-source-collection-useful-apache-pig-udfs.

In addition to these Pig statements, the following R UDF is defined in the RFuncs.r script.

```
Quantile.outputSchema ← "q:double";
Quantile ← function (x, probs, type) {
    probs ← as.numeric(unlist(strsplit(probs, split = ",")))
                            # parse the parameter value
    q ← quantile(unlist(x), probs = probs, names = T, type = type);
                            # call the R quantile() function
    return (as.list(q)); }
```

The *Quantile* UDF is a simple wrapper for the quantile() function of the R stats library. x is a numeric vector whose sample quantiles are desired. Its value is converted from the Pig input tuple by the framework. The function parameters (probs, type) value can be supplied in different ways, for example, a Declare statement that is used in the example or a Parameter File and so on.

To summarize this use case of RPig, any original R function can be easily wrapped and exposed as a Pig R UDF. The necessary input parameter of the original R function can be exposed by the UDF to make the function more generic for reusability. Still, all the input data for a single function call will be executed in one R engine, and some partitioning might be necessary (e.g., group by "week") if the data are too large.

9.5.1.2 Result and Discussion

In this use case of computing quantiles, both DataFu and RPig only require a few lines of Pig (and R) code as the user does not need to write the quantile algorithm. However, the RPig implementation of the function is much more flexible regarding the data input and output formats than DataFu. The DataFu quantile function only takes a sorted input bag, and each numeric value is a tuple inside the bag. We have to preformat the raw data before calling the function in this case. In contrast, the RPig version can handle any format of bag or tuple input. Numeric values can be either in one tuple/bag or separated tuple/bags since the data always will be flattened into numeric vectors in the R function before computing quantiles.

Figure 9.2 shows the performance comparison with a fixed 20-node Hadoop cluster. Each row of input data contains more than 10K double values for one network node, and that makes around 1 GB raw data for every 10K rows. The RPig version implementation with the JVM R engine (Renjin 0.7.0) has the slowest performance. It becomes very slow when input data size becomes larger, and it consumes almost all available memory for the map task. It might relate to the internal memory management problem of Renjin since it is only in a very early stage. The RPig with stand-alone R has the best performance. DataFu (v 0.0.10) is in the second since it needs to preformat and sort the data

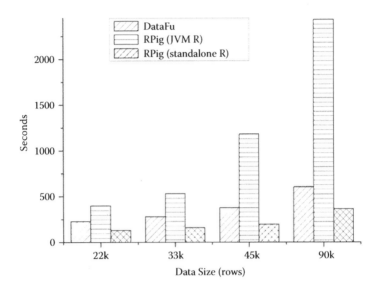

FIGURE 9.2
The performance comparison on DataFu and RPig.

through Pig operations, and these take more time before calling the quantile function.

DataFu has some convenience bag (e.g., enumerating bags) and utility functions, but the availability of statistical functions in DataFu is extremely limited. It only includes common statistics tasks (e.g., quantile, variance), PageRank, and the like algorithms that are relevant to the LinkedIn use cases. Even for the quantile function, DataFu only implements the type R-2 estimation, which is one of several algorithms for estimating quantiles. RPig allows the use of nine quantile algorithms implemented in R, selected by the `type` parameter in the example. With RPig, it is easy to wrap and expose any statistical function of R as a Pig UDF. The statistical functions available in RPig are as many and as comprehensive as in the original R.

In summary, RPig provides extensive statistical and machine learning algorithms by wrapping any original R function in a Pig UDF, and the UDF is flexible with input and output data formats and gives the best performance (with stand-alone R) in the this case. In contrast, DataFu is ready to use without needing additional installation of a script engine since it runs on the JVM, but the number of functions is extremely limited.

9.5.2 Forecasting with EMA

9.5.2.1 Design and Implementation

EMA is used for forecasting data traffic on selected VoIP service clients for a use case described in Section 9.2. Since EMA is a light algorithm, and the aggregated data `Traffics` (from Section 9.3.3) is already small enough in this case, we can group all the data together and send them to one R engine using RPig. The following shows the Pig statements:

```
Results = FOREACH (GROUP Traffics ALL) GENERATE RFuncs.ema_all
   ($1, n);
```

ema_all() is a defined R function processing the grouped data, as in the following:

```
ema_all.outputSchema ← toString(lapply(seq(1,11), function(x)
   {paste("map[tuple(double)]", sep = "")}))
ema_all ← function(x, n) {
  xDf ← as.data.frame(do.call(rbind, x[[1]]))
                        # convert to a data frame
  ...                   # sorted the data and initial variables
  library('TTR')
  for(i in 1:length(clients)){
     t ← xDf[xDf[,c(3)] = =clients[i], c(4)]
     results ← append(results,list(list(as.character(clients[i]),
       EMA(t,n))))}
return (results)}
```

In this case, the data passed to R is a nested list (x), which contains aggregated traffic data for all service clients in different time windows, $\left(\left(x_1^1, x_2^1, x_3^1 \right), \left(x_1^2, x_2^2, x_3^2 \right), \dots \right)$. The first line of the R script converts the nested list to a data frame called xDf, so the input data can be easily sorted and selected as a data table. A sorted numeric list containing traffic data of previous time windows for each service client is selected and is used as input for the R `EMA()` function of the TTR package. Results of all service clients as a nested list `results` will be subsequently converted to a Pig map data structure specified by the output schema. The name of the service client is the key of the map, and the forecasted result is the value of the map. In this case, the Pig statement is used as the query language for accessing the data from the HDFS file system, and then the converted data will be sent to R for analytic tasks. Afterward, the data analytic result is printed on screen or stored in HDFS through Pig statements. Hence, RPig can be used as a way for R programmers to read and write data and files in HDFS.

To summarize this use case of RPig, the Pig operations are used as preprocessing steps to extract and summarize only the necessary information needed for R processing. When the summarized data are small enough to be handled in R in a single node, then we can use any statistical algorithm implementations of R directly on the summarized data similar to the traditional single-machine approach of R.

9.5.2.2 Result and Discussion

The necessary data must be converted and loaded into R first when an R function is involved in a Pig data flow, and we consider this as performance overhead. Minimizing this overhead was one of our main tasks after the initial version of RPig development. As a consequence, the initial version of RPig was used for a comparison study in this case. The code implementation for this use case based on the initial version can be found in Reference 9, and it is similar to the implementation using the current version.

Figure 9.3a shows the results with 20 fixed nodes. The data size represents the initial raw data size loaded in Pig. With both versions of the implementations (both with stand-alone R engines), the performance decreased with increasing data size, as expected. In this scenario, the performance mainly depends on Pig/Hadoop, which needs to handle a large amount of raw data, where R only plays a small part in the overall process. We can see the current version has better overall performance, and the improvement becomes larger when more data are involved. Figure 9.3b shows the improvement in detail when sending data from Pig to R in a single node. In this case, summarized data with more than a half million data tuples and four data fields in each tuple will take 20 seconds in the initial version but only takes 10 seconds in current version. Overhead is reduced 50% in the current version. This is achieved by sending data directly to R through the socket connection and many code optimizations in the current version. The initial version of RPig streams the data to the disk as an R source file, then makes R load the source file. Still, when more data need to be exchanged between R and Pig, then the overhead becomes larger. This overhead can be considered as a trade-off between user development effort and processing efficiency. We only have 10 lines of R code in the R functions in this use case, but we or the user had to write around 100 lines of Java code for the EMA Pig UDF without using

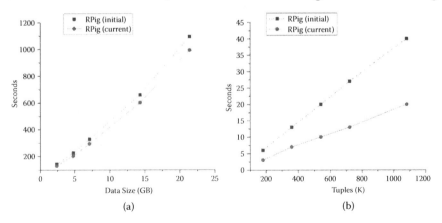

(a)   (b)

FIGURE 9.3
(a) Overall performance comparison. (b) Overhead comparison on data exchange (Pig to R).

R for the same calculation. In another example, DataFu has around 200 lines of Java code for the Quantile function in the first use case. This shows the significant reduction of coding and code maintenance effort with RPig.

In summary, RPig offers concise programming for data analytics by utilizing existing implementations of algorithms in R. However, the necessary data required must be converted and loaded into R; this causes the performance overhead. As a result, data should always be minimized as much as possible before exchanging data between Pig and R to reduce overhead. The current RPig has a significant performance improvement over the initial implementation; it has reduced the overhead of data exchange by 50%.

9.5.3 Prediction with SVM

9.5.3.1 Design and Implementation

Because an SVM model is constructed based on determined support vectors in SVM algorithms, an SVM training data set can be represented by data samples as support vectors. The remaining data of the data set that do not directly contribute to the final SVM model can be viewed as redundant, even though minor inaccuracies may occur in some cases [13, 14]. Therefore, if we have two map functions, one (map_{sv}) is for extracting samples marked as support vectors from a data set, another (map_{svm_m}) is for having an SVM model from a data set, and a generic reduce function (*reduce*) is only to aggregate a list of results from map functions, then the SVM training phase to obtain a model for a data set D can be defined in the MapReduce model as the following.

```
Training Phase:
repeat a number of times if required:
    split D to {D₁,D₂,...,Dₙ}
    D ← in parallel execution: reduce(map_sv(D₁),map_sv(D₂),...map_sv(Dₙ))
Model ← reduce(map_svm_m(D))
```

Since support vectors are often only a small data subset of the original input data set $map_{sv}(D) < D$, and map and reduce functions are executed in parallel in Hadoop, building an SVM model from a data subset would be much faster than building the original data set. Hence, the overall SVM training is expected to be scalable with the size of the cluster. A parallel algorithm can also be structured as multiple rounds of map and reduce. Collected samples as support vectors can be treated as a new data set; hence, the map_{sv} can also be applied repeatedly to further reduce the data size if it is required.

In the prediction phase, it takes the trained model and network KPIs at the IP layer, such as packet loss, as input, then gives a predicted MOS value instantly. In this case, we want to do MOS value prediction in parallel for a large amount of VoIP call sessions S, then a map function $map_{predict}$ can be defined to take a subset of call sessions to increase the scalability.

Training Phase:

```
split S to {S₁,S₂,...,Sₙ}
results ← in parallel execution:
   reduce((map_predict(Model,S₁),map_predict(Model,S₂),...,
       map_predict(Model,S₂))
```

As analyzed, we have the following script for our scenario: We first describe the training phase implementation. The following code fragment shows the step of extracting support vectors from a split data set StatisticEvents _ s in the MapReduce model:

```
SV = FOREACH StatisticEvents_s GENERATE FLATTEN(RFuncs.svm_sv($1));
   svm_sv.outputSchema ← "bag{tuple(double, double)}";
   svm_sv ← function(x) {
      xDf ← as.data.frame(do.call(rbind, x[[1]]))
                              # data frame of training dataset
      ...                     # extracting the support vector sv
      return (list(sv))
   }
```

The R function svm _ sv is an implementation of the map map$_{sv}$ function. Extracting support vectors is the same as building an SVM model. It covers cross validation, parameter tuning, and so on for complete SVM training. However, instead of obtaining a final SVM model, we only fetch out the samples as support vectors after the SVM training. In our case, using a radial-kernel-based SVM regression, SVM computation can be represented to solve the following optimization problem:

$$\min_{\alpha,\alpha^*} \frac{1}{2}\left(\alpha-\alpha^*\right)^T y_i y_j \exp\left(-\gamma\|x_i-x_j\|^2\right)\left(\alpha-\alpha^*\right)+\varepsilon\sum_{i=1}^{n}\left(\alpha-\alpha^*\right)+\sum_{i=1}^{n}y_i\left(\alpha-\alpha_i^*\right)$$

$$\text{subject to } 0\le\alpha_i,\alpha_i^*\le C,\sim i=1,...,n,\sum_{i=1}^{n}\left(\alpha-\alpha_i^*\right)=0\sim.$$

where x_i, x_j are the input data set, *gamma* is a parameter. In R, we use the *e1071* package to supply the SVM implementation mentioned. We first use a tuning function to find the best parameter over a parameter range, and then we train an SVM. The following R code fragment shows the part of extracting *sv*:

```
tuned ← tune.svm(V2 ~., data = xDf, gamma = 10^(-2:2),
   kernel = 'radial') # turn the parameter
svmModel <- svm(xDf[2], xDf[1], kernel = 'radial',
   gamma = tuned$best.parameters$gamma, cross = 10)
sv ← xDf[c(svmModel$index),]
sv ← apply(sv, 1, function(x){as.list(x)})
                              # put each row to a list
```

Here, sv is extracted samples as support vectors from the input date frame xDf. However, the data frame of R is a column-oriented structure. All of the values of a column are grouped together, then the values of the next column are in a second group, and so on. Data tables stored in Hadoop and Pig are the same as the commonly used CSV (comma-separated value) format, which is primarily row oriented. If we want to use the Pig SPLIT operator to split the collected support vectors to repeat the sv extracting process again, we need to convert the collected data representation to be row oriented. Hence, we use the apply() function to put each row to a list. Finally, all lists will be put as tuples into a bag sent back to Pig. We flatten the bag in Pig and convert the data back to the "table" format to continue processing.

In the last step of the training phase, we group the finalized data sets or support vectors *SV* and send them to one R engine to obtain a final SVM model and then store the model for the prediction phase. The MapReduce model is still applied, but only one map and one reduce function will be created at this stage.

```
Model = FOREACH (GROUP SV ALL) GENERATE R.svm_m(*);
svm_m.outputSchema ← "model:bytearray";
svm_m ← function(x) {
    ...                                    # get the svmModel
    return (serialize(svmModel, NULL)) }
```

The R UDF svm_m is almost the same as the previous svm_sv but returns an SVM model svmModel this time. The serialized model will be saved as a bytearray or original R object in HDFS, so we can use the model directly in R for prediction later. In the SVM prediction phase, the SVM model can be directly loaded into R from Pig in parallel execution for a huge number of VoIP call sessions.

RHadoop [3] is a popular open-source project from Revolution Analytics that allows users to manage and analyze data with Hadoop in R. The rmr2 is an R package from RHadoop; it offers the user the ability to write MapReduce functions in R. We implemented parallel SVM design with rmr2 for the comparative study in this case. The following shows the MapReduce implementation to obtain support vectors *SV*: We use this function as the example to show the difference in implementation with regard to the RPig version. The *SV* has exactly the same value as the RPig implementation we described.

```
svDfs ← mapreduce(input = inputPath,
    map = function(dummy, input) {
        ...                        # extracting the support vector sv
        keyval(1, list(sv)) },
    reduce = function(k, sv){
        val ← do.call("rbind", sv); keyval(1, val) }
)
SV ←from.dfs(svDfs)$val
```

The map function extracts the support vectors of every data subset and outputs a key-value pair, the value part is the support vectors sv. All outputs of the map functions have the same key value, integer 1, so the extracted sv of different data subsets will be collected and aggregated together by the reduce function. The final result SV can be retrieved for the value part of the key-value pair output of the reduce function.

To summarize this use case with RPig, parallel or iterative statistical algorithms for distributed data sources are expressed as parallel R executions in a Pig data flow. Input data are treated as a number of distributed data sources with no centralized information during parallel R executions for each data source; aggregated results of distributed R executions as stepping stones are relative to a final result of a final centralized R execution. Pig operations are used to distribute the data and tasks for parallel processing with multiple R engines as Map tasks. This approach allows parallel R executions to reduce the processing time. However, statistical errors may be caused by the iterative and incremental statistical algorithms as a trade-off and are acceptable in most cases [13, 14].

9.5.3.2 Result and Discussion

Both RPig and RHadoop allow a parallel SVM implementation in the MapReduce model. RPig just uses the FOREACH statement to parallelize the tasks as the Map functions. RHadoop allows the user to code the entire analytic job in R, but the user has to design the key-value pairs based on Map and Reduce functions. This creates complexity for the user in code design and development compared to RPig, especially when multiple MapReduce functions are necessary for complex analytic jobs. In the example described for obtaining SV, we wrote 16 lines of Pig and R code using RPig, but needed 21 lines of R code for RHadoop because of writing the key-value pair functions (Figure 9.4a). This again shows the concise programming of RPig.

A relatively small size data is used in this case since it is very CPU intensive as described previously. We split the data containing 12K training samples into 16 pieces to obtain support vectors and then we obtained the final SVM model at the end of the training. The performance of the SVM training phase with respect to the cluster size is shown in Figure 9.4b. RPig has almost identical performance compared to the RHadoop (rmr2 v2.2.2) implementations. There is a reduction of processing time as the cluster size increases, but the decrease in processing time is not exactly linear as there is a higher communication cost with a larger cluster.

In summary, RPig is able to scale out machine learning functionalities for deep analytics. We demonstrated this through an SVM use case. RPig is less complex to use and requires less development effort for writing parallel machine learning algorithms compared to RHadoop or others (e.g., RHIPE [15]), which require designing and writing key-value-paired Map and Reduce functions manually.

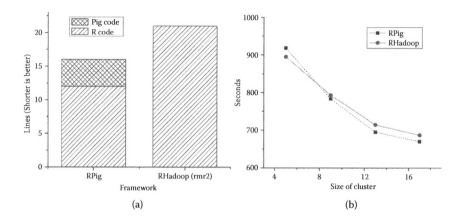

FIGURE 9.4
(a) Lines of code comparison. (b) Scalability on SVM training.

9.6 Related Work

9.6.1 Related to R

With the emergence of big data analytics, many researchers are addressing the scalability issues of R. The existing approaches can be classified in three different categories:

1. *Scaling memory size*: All data used in R calculations, such as lists and data frames, need to be loaded into memory; however, a single computer only has a limited memory size that restricts a large data set from being loaded into R. RevoScaleR [16] and bigmemory [17] are R packages that allow R to use a hard disk as external memory for calculations. This approach allows R to handle a data size much bigger than its memory size since the size of the hard disk is generally much larger than the memory size of a computer.

2. *Scaling storage size*: Terabyte-level big data are generally stored in distributed file systems, such as Hadoop clusters. To enable R to directly read/write data in these large-scale data warehouses, interfaces between these warehouses and R are developed, such as Ricardo [18], which offers a bridge between R and Hadoop HDFS. Comparing Ricardo to R bridging work on traditional RDBMS such as RJDBC [19] and RMySQL [20], SQL is replaced by a query language (Jaql), which can be executed in the MapReduce model in Ricardo. These approaches allow R to directly access data from database or file systems, but the R script execution remains in a single computer. For parallel data analysis, it requires reimplementing most of a statistic algorithm in the

query language. In other words, we have to reimplement the SVM()
function of R in Jaql for our second use case to parallelize the process.

3. *Scaling CPU power*: Approaches for scaling out CPU power for R
can be divided into MapReduce- and non-MapReduce-based implementations. MapReduce-based approaches are generally running
on top of Hadoop. For example, both RHIPE [15] and RHadoop [3]
extend R to allow writing key-value pair map and reduce functions
within an R script. The MapReduce jobs of R can be submitted to a
Hadoop cluster for parallel executions. However, these frameworks
require users to manually design complex key-value pair-based map
and reduce functions, making them difficult to use and inefficient
for analysis job development. In our case with RPig, the key-value
pair-based map and reduce functions are automatically generated
by leveraging the Pig framework. The user only needs to define R
functions for a single task node; the execution of the R functions is
parallelized automatically based on Pig data flows. RHive [21] has
the same concept as our work. It is an R extension facilitating distributed computing via HiveQL/SQL queries. However, it is restricted
for the Hive data warehouse. And, considering the natural differences between Pig and SQL language, RHive is an alternative to
RPig, but it cannot be a replacement.

Many approaches utilize non-MapReduce-based parallel frameworks,
such as Open MPI [22], and packages such as Rmpi [23] and snow [24] provide bridge interfaces between R and MPI. CloudRmpi [25] supports management of an EC2 cluster and access to an R session on the master MPI node.
Elastic-R [26] allows users to send data to any R engine in an R engine pool.
However, the solutions do not support parallel data read/write as Hadoop;
hence, they are not suitable for I/O-intensive scenarios. Furthermore, these
solutions are difficult to use as the user must code send/receive message
functions for master and slave nodes through complex MPI API.

9.6.2 Other Related Solutions

Some approaches try to build new systems without using traditional statistical
frameworks, such as R. For example, Mahout [2] is a framework built on top
of Hadoop with MapReduce-based machine learning algorithms. However,
Mahout is only at an early stage; many commonly used algorithms, such as
SVM, are not available yet. Second, it does not provide a high-level language,
such as R and Pig; instead, complex Java APIs are provided. As a result,
developing analytic jobs in Mahout is complex and difficult. SystemML [27]
proposes a new declarative machine learning (DML) language for machine
learning on MapReduce. However, DML is not as flexible as R language;
it does not support object-oriented features, advanced data types (such as
lists and arrays), and so on in comparison with R. More important, SystemML

is the same as other newly developed frameworks, such as MLbase [28] and Cloudera ML [29], and also lacks commonly used statistic and machine learning algorithm implementations.

9.7 Conclusion

R provides comprehensive machine learning and statistical algorithms. However, the R execution environment is not distributed and is not considered scalable. In contrast, Pig supports parallel data processing using high-level language; but it does not provide implementations of common statistical algorithms; it lacks the necessary features for advanced statistical analysis. In this chapter, we presented an integrated RPig framework that takes advantage of both R and Pig, allowing scalable deep analysis while minimizing the development effort with concise programming.

We have described the design and implementation of an RPig framework. Based on the use case scenarios, we have demonstrated the use of our framework. We have shown experimental results related to scalability and coding effort reduction with examples. We also did a comparison study in each use case experiment to show the difference or improvement over related work. Our future work will create an R package that would allow calling Pig in R.

References

1. Dean, J., and S. Ghemawat, MapReduce: simplified data processing on large clusters. *Communications of the ACM*, 2008, 51(1): 107–113.
2. Apache Mahout. Home page. http://mahout.apache.org/.
3. RHadoop. https://github.com/RevolutionAnalytics/RHadoop/wiki/.
4. Handurukande, S., et al. Magneto approach to QoS monitoring. In *IFIP/IEEE International Symposium on Integrated Network Management*. 2011.
5. White, T., *Hadoop: The Definitive Guide*. 2nd ed. Sebastopol, CA: O'Reilly Media, 2011.
6. Eaton, C., et al., *Understanding Big Data: Analytics for Enterprise Class Hadoop and Streaming Data*. New York: McGraw-Hill, 2012.
7. DataFu. http://data.linkedin.com/opensource/datafu.
8. Olston, C., et al. Pig Latin: a not-so-foreign language for data processing. In *ACM International Conference on Management of Data*. 2008.
9. Wang, M., S. B. Handurukande, and M. Nassar. RPig: a scalable framework for machine learning and advanced statistical functionalities. In *IEEE International Conference on Cloud Computing Technology and Science*. New York: IEEE, 2012.
10. Renjin. Home page. http://www.renjin.org/.
11. rsession. http://code.google.com/p/rsession/.

12. Chiba, S., and M. Nishizawa. An easy-to-use toolkit for efficient Java Bytecode translators. In *International Conference on Generative Programming and Component Engineering*. 2003.
13. Syed, b.N.A., et al. Incremental learning with support vector machines. In *International Knowledge Discovery and Data Mining Conference*. 1999.
14. Graf, H. P., et al., Parallel support vector machines: the cascade SVM. *Advances in Neural Information Processing Systems*, 2004, 17: 521–528.
15. Guha, S., Computing Environment for the Statistical Analysis of Large and Complex Data, doctoral dissertation, Purdue University, 2010.
16. Rickert, J., *Big Data Analysis with Revolution R Enterprise*. White paper. Mountain View, CA: Revolution Analytics, 2011.
17. bigmemory. http://cran.r-project.org/web/packages/bigmemory/index.html.
18. Das, S., et al. Ricardo: integrating R and Hadoop. In *ACM International conference on Management of Data*. 2010.
19. RJDBC. http://www.rforge.net/RJDBC/index.html.
20. RMySQL. http://cran.r-project.org/web/packages/RMySQL/.
21. RHive. Available from: https://github.com/nexr/RHive.
22. Open MPI. http://www.open-mpi.org/.
23. Rmpi. http://www.stats.uwo.ca/faculty/yu/Rmpi/.
24. snow. http://cran.r-project.org/web/packages/snow/index.html.
25. cloudRmpi. http://norbl.com/cloudrmpi/cloudRmpi.html.
26. Chine, K., Open science in the cloud: towards a universal platform for scientific and statistical computing. In *Handbook of Cloud Computing*, B. Furht and A. Escalante, editors. New York: Springer, 2010, pp. 453–474.
27. Ghoting, A., et al. SystemML: declarative machine learning on MapReduce. In *IEEE International Conference on Data Engineering*. 2011.
28. Kraska, T., et al., MLbase: A Distributed Machine Learning System. In *The Conference on Innovative Data Systems Research*. 2013.
29. Cloudera ML. https://github.com/cloudera/ml.

10

AutoDock Gateway for Molecular Docking Simulations in Cloud Systems

Zoltán Farkas, Péter Kacsuk, Tamás Kiss, Péter Borsody,
Ákos Hajnal, Ákos Balaskó, and Krisztián Karóczkai

CONTENTS

10.1 Introduction

Parameter sweep applications are frequent in scientific simulations and in other types of scientific applications. They require running the same application with a very large number of parameters; hence, their execution time could be long on a single computing resource. As a result, collecting resources on demand from distributed computing infrastructures (DCIs), such as clouds or grids, is a highly desired feature of any computing environment that is offered for scientists to run such applications. Cloud computing infrastructures are especially suitable for such applications due to their elasticity and easy scaling up on demand.

Molecular docking simulations are a widely utilized application area in which parameter sweep scenarios are desired. Molecular docking simulation packages, for example, AutoDock [1], are applied by various disciplines, such as molecular biology, computational chemistry, or even psychology, and require a large number of cloud resources to increase the speed of computation.

To collect the required number of cloud resources, end users like biologists and chemists would have to learn the cloud interfaces. However, instead of learning such information technology (IT) systems, they would rather like to concentrate on their own scientific field and research. To hide this low-level technology from them, high-level user interfaces like science gateways are required. WS-PGRADE [2] was designed with this idea to provide high-level graphical workflow abstraction for users to hide the low-level details of accessing the underlying cloud infrastructures. WS-PGRADE provides workflow templates that tremendously simplify the creation of parameter sweep (and other types of workflow) applications and takes care of accessing the required type and number of cloud resources. To achieve this, WS-PGRADE was integrated with the CloudBroker Platform (CBP), which enables accessing heterogeneous cloud resources, including Amazon EC2, IBM SmartCloud, Eucalyptus [3], OpenStack [4], and OpenNebula [5] clouds. Moreover, WS-PGRADE also supports the development of intuitive and customized end-user interfaces to completely hide the underlying complexity from the scientist.

This chapter demonstrates how parameter sweep application scenarios, such as AutoDock-based molecular docking experiments, on cloud computing infrastructures can be efficiently supported by the WS-PGRADE framework that is the core technology of the European Union FP7 project SCI-BUS [6], which is aimed at developing various science gateways for a large set of different scientific user communities.

10.2 WS-PGRADE Workflows and Parameter Sweep Application

WS-PGRADE workflows are represented as directed acyclic graphs (DAGs), for which nodes denote computational tasks (or some other workflow), and directed arcs denote data dependency between the different nodes. Figure 10.1 shows a workflow with six nodes, where different data dependencies are defined. When submitting this workflow, the first two nodes (Copy_A and Copy_B), as they have no prerequisites, are able to run immediately on the targeted computing infrastructure. Once Copy_A has finished and produced the necessary input for the Invert_A node, this job is ready for submission. Similarly, once Copy_B and Invert_A have both finished, WS-PGRADE can start processing the Multi_B node. That is, once all the preceding nodes of a given workflow node have finished successfully, the given node is run by WS-PGRADE.

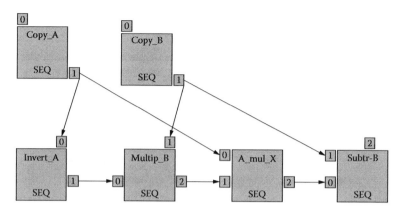

FIGURE 10.1
Sample WS-PGRADE workflow.

In a nonparametric workflow, every node is run only once. However, it is possible to construct parameter sweep workflows in which a node of the workflow is submitted multiple times using different input data for the different submissions. Figure 10.2 shows an example construct for a parameter sweep workflow.

Figure 10.2 has been split into three parts: the generator phase at the top, the processing (parameter sweep) phase in the middle, and the resulting collecting phase at the bottom. In the generator phase, WS-PGRADE runs special nodes in the workflow, the generator nodes. The task of these nodes is to produce the parameter space for the computation, for example, by splitting one big input data set into smaller chunks. The generator nodes produce the different inputs for the actual computation. In the processing phase, the nodes process all the inputs created by the generator nodes. If there were multiple

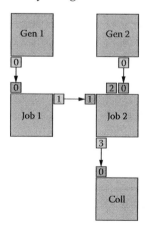

FIGURE 10.2
Sample parameter sweep WS-PGRADE workflow.

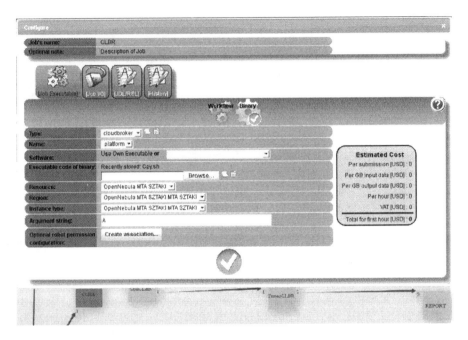

FIGURE 10.3
Configuration of a WS-PGRADE workflow node.

generator nodes (as shown in Figure 10.2), then it is possible to specify how to pair the different inputs, using either cross or dot products of the files produced by the different generator nodes. Finally, if all of the node instances have finished in the processing phase, the collector nodes receive the results of the computations and can process them (e.g., they can search for the best result or can produce some sort of statistics). Note that these phases can be overlapped or repeated within a workflow; that is, generators and collectors can be placed into a workflow at any part of the graph without restrictions.

The configuration of workflow nodes can be performed through a simple interface, as shown in Figure 10.3. It is possible to define the execution resource, the application, different resource-specific settings, the command line arguments, and the data used and produced by the job.

Once the configuration of the whole workflow has finished, it can be submitted. From this point, WS-PGRADE/gUSE is responsible for taking care of the nodes' execution.

10.3 AutoDock Workflow

This section explains how to use the generic parameter sweep creation and execution technology described in the previous section for a concrete application, namely, AutoDock. In fact, three different variants of AutoDock parameter sweep workflows have been developed, and they are described.

10.3.1 The AutoDock Application

Traditionally, in vitro studies have been used to investigate the binding of receptors to their ligands and enzymes to their substrates. These wet laboratory experiments are both time consuming and expensive to carry out. An in-silico system based on computer simulations can facilitate the modeling of receptor-ligand interactions prior to the wet laboratory experiments and enable scientists to better focus the in-vitro experiments using only the most promising molecules. With the advances in computer technology, it is now feasible to screen hundreds of thousands of compounds against a single receptor to identify new inhibitors or therapeutic agents. However, the modeling programs are not user friendly, and the relationship between results obtained by molecular modeling and by in-vitro studies and the newer biosensors still needs to be established.

AutoDock is one example of a program that allows in-silico modeling of intermolecular interactions. AutoDock is a suite of automated docking tools. It is designed to predict how small molecules, such as substrates or drug candidates, bind to a receptor of known three-dimensional (3D) structure. AutoDock currently comprises two discrete generations of software: AutoDock4 and AutoDock Vina.

AutoDock 4 is typically used to accurately model the molecular docking of a single ligand to a single receptor. In this instance, the process is composed of three discrete stages. First, a low-complexity sequential preprocessing stage defines a random starting location in 3D space (termed the *docking space*) for both the ligand and the receptor. This is achieved using a tool within AutoDockTools (ADT) called AutoGrid. The locations, which are characterized by atomic energy levels at each point within the docking space, act as a single common input to a second stage. The second stage can comprise many parallel jobs, each receiving a copy of the ligand and receptor starting locations that form the input to a genetic algorithm. The algorithm acts to randomly rotate/reposition the ligand and then determine likely docking/binding sites based on energy levels, which are calculated from the original starting locations. This process can be considered a parameter sweep, where the varied input parameter is the initial random rotation of the ligand. Finally, a single low-complexity sequential postprocessing stage can be used to identify the most likely binding sites by comparing energies from all jobs of the preceding stage (where minimized energies represent likely docking sites).

AutoDock Vina provides several enhancements over AutoDock4, increasing average simulation accuracy while also being up to two orders of magnitude faster. Autodock Vina is particularly useful for virtual screening, whereby a large set of ligands can be compared for docking suitability with a single receptor. In this instance, parallelism is achieved by first breaking the set of all ligands into equal-size disjoint subsets. Each compute job then uses a different subset as an input. The ligands in each subset are simulated/docked sequentially on the compute node using the single receptor; a postprocessing stage can be used to compare the results from all compute jobs.

Researchers from the School of Life Sciences at the University of Westminster in London have set up a novel screening system [7] to analyze well-characterized protein-ligand interactions, for example, studying the interrogation of enzymes and receptors of the protozoan *Trichomonas vaginalis* (TV). TV is an important organism, with 180,000 million women affected worldwide. It is also a proven cofactor for the acquisition of human immunodeficiency virus (HIV). Currently, only one drug is available, metronidazole, and resistance has been reported. The cloning and publication of the TV genome offers new options for drug/inhibitor detection utilizing bioinformatics and molecular modeling tools.

Westminster researchers constructed an in silico small molecule library of about 300,000 structures. Given a receptor file and the approximated position and size for the active site, the whole library was planned to be screened against the chosen receptor using the AutoDock Vina (http://vina.scripps.edu/) molecular docking tool. Once operational, the system could easily be utilized for other similar virtual screening experiments.

10.3.2 AutoDock Workflows

Three different parameter sweep workflows were developed (in the framework of the EU FP7 ER-Flow project [8]) based on the AutoDock4 and AutoDock Vina applications and the previously described scenarios: the AutoDock workflow, the AutoDock without AutoGrid workflow, and the AutoDock Vina workflow.

The AutoDock workflow requires PDB (Program Database) input files (these are widely available in public databases), automatically converts these files into PDBQT format (which is required by the AutoDock application), calculates the docking space running the AutoGrid application, and docks a small ligand molecule on a larger receptor molecule structure in a Monte Carlo simulation. Finally, it returns the required number of lowest-energy-level solutions. The workflow uses version 4.2.3 of the AutoDock docking simulation package. Users of this workflow are expected to provide input files for AutoGrid (molecules in PDB format), grid parameter file (GPF), docking parameter file (DPF), the number of simulations to be carried out, and the number of required results. The workflow is shown in Figure 10.4. This workflow is ideal for researchers who are less familiar with the AutoDock suite and command line tools and require a high level of automation when executing their experiments.

On the other hand, the AutoDock without AutoGrid workflow requires the scientist to run scripts from the AutoGrid application on his or her own

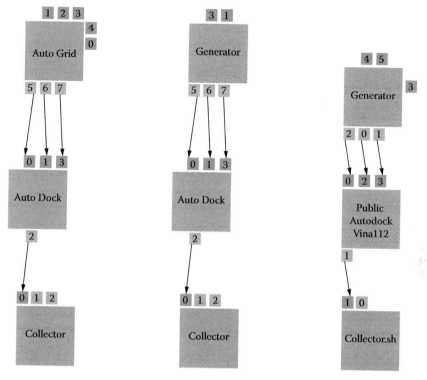

FIGURE 10.4
AutoDock workflow.

FIGURE 10.5
AutoDock workflow without AutoGrid.

FIGURE 10.6
AutoDock Vina workflow.

computer prior to the execution of the workflow. Although this requires specific expertise, it also gives much more flexibility to the end user when preparing the input molecule and the docking space. As a consequence, this workflow requires PDBQT input files and the output of the AutoGrid application, and (similar to the previous workflow) it docks a small ligand molecule on a larger receptor molecule structure in a Monte Carlo simulation using 4.2.3 of the AutoDock docking simulation package. Users of this workflow are expected to provide a docking parameter file, a ZIP file containing input files for AutoDock that were generated using version 4.2.3 of AutoGrid and the former docking parameter file, the number of simulations to be carried out, and the number of required results. This workflow is shown in Figure 10.5.

Finally, the AutoDock Vina workflow performs virtual screening of molecules using version 1.1.2 of AutoDock Vina. It docks a library of small ligands on a selected receptor molecule. Users of this workflow are expected to provide a configuration file for AutoDock Vina, an input receptor molecule in PDBQT file format, a ZIP file containing a number of ligands in PDBQT file format, the number of simulations to carry out, and the number of required results. The workflow is shown in Figure 10.6.

As can be seen in the figures, all of the workflows follow the generator node–parameter sweep node–collector node semantics. The generator nodes are executed locally, on the portal server. The parameter sweep nodes are executed in a targeted distributed computing resource, which in the first version of the gateway was the EDGeS@home volunteer desktop grid (DG). It is important to note that, for performance optimization, the AutoDock and AutoDock without AutoGrid workflows are submitting one single metajob to the desktop grid server. This means that the actual parameter sweep expansion will happen on the DG server and not on the portal side (i.e., from the portal's point of view, these workflows are not real parameter sweep workflows, but from the point of view of the whole processing, they are, as a number of workflow node instances will be generated on the DG server). The collector nodes are executed locally, on the portal server.

The task of the generator nodes is to set up the parameter space for the middle nodes. In the case of the AutoDock and AutoDock without AutoGrid workflows, this node is simply generating a metajob description for the DG server, whereas in the case of the AutoDock Vina workflow, the generator node distributes the ligands provided by the user in a ZIP file into as many packages as the number of simulations set by the user during the workflow's configuration.

The middle node of the workflows is responsible for the actual parameter sweep processing. As mentioned, in the case of the AutoDock and AutoDock without AutoGrid workflows, the parameter sweep expansion happens on the DG server, whereas in the case of the AutoDock Vina workflow, this happens on the portal server, and parameter sweep job instances will be submitted to the DG server.

Finally, the task of the collector nodes is to evaluate the results of the parameter sweep executions and collect the best dockings for the user.

10.4 Migrating the AutoDock Workflows to the Cloud

This section presents how the AutoDock workflows have been migrated to a cloud infrastructure, including all the necessary features of WS-PGRADE/gUSE and the migration process.

10.4.1 Cloud-Based Execution in WS-PGRADE

WS-PGRADE offers access to a number of DCIs, including clusters, grids, DGs, and clouds. Access to cloud systems is solved with the help of the CBP [9]. The CBP offers unified access to most major cloud infrastructures, such as Amazon EC2, IBM, Eucalyptus, OpenStack, and OpenNebula, at three different levels: the web interface, a RESTful web service, and a Java application programming interface (API). The web interface and the REST API offer access to most of the CBP functionalities, with the first one offered for users and the second one offered for developers for integrating CBP features into their products. Finally, the Java API offers a convenient tool for accessing the majority of CBP features from Java applications. WS-PGRADE uses the Java API of CBP to access the different cloud services.

The integration of WS-PGRADE and CBP aims to hide details of the cloud infrastructure used. As shown in Figure 10.3, after the users have selected to use a cloud infrastructure for their workflow node, all they have to do is select an application already deployed in the cloud (or indicate that they would like to run their own application and upload it) and select a cloud resource for the computation, for example, Amazon EC2 or OpenNebula. Once the workflow has been configured and submitted, execution of the selected applications with the provided data is arranged in the background by WS-PGRADE/gUSE and the CBP.

The integration architecture of WS-PGRADE/gUSE and CBP is shown in Figure 10.7.

The top of Figure 10.7 represents WS-PGRADE and gUSE. Based on WS-PGRADE, a number of customized science gateways (Proteomic, Seismology, Rendering, etc.) can be created that can hide the workflow concept of WS-PGRADE through a simplified user interface. WS-PGRADE

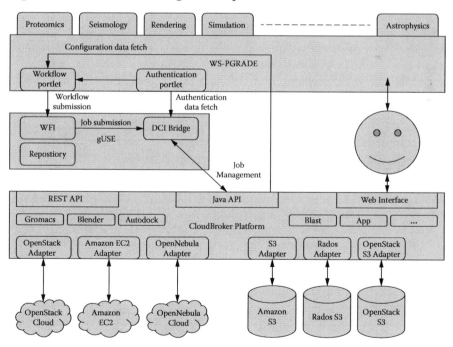

FIGURE 10.7
Architecture of the WS-PGRADE/gUSE and CBP integration.

itself interacts with the CBP through two different portlets: the Workflow portlet (for creating, configuring, and running workflows) and the Authentication portlet (for specifying the CBP credentials to be used by the user). The CBP credential set by the user in the Authentication portlet is used by WS-PGRADE/gUSE to communicate with the CBP service on behalf of the user (please note that the requirement toward users to specify CBP credentials can be eliminated by assigning robot certificates to existing workflows).

Once the workflow has been configured and submitted with the help of the Workflow portlet, the set of back-end components (gUSE) are responsible for arranging the workflow's execution. The Workflow Interpreter (WFI) is used to schedule nodes of the workflow for execution, and the DCI Bridge is used to actually make the different job instances run in the selected DCI (cloud, in our case). Both WS-PGRADE and gUSE components are using the CBP Java API to access the CBP service. Once the individual job instances have been sent to the CBP by the DCI Bridge, the CBP is responsible for arranging the jobs' execution on the selected cloud service.

10.4.2 Robot Certificates in WS-PGRADE/gUSE

As presented previously, accessing the services of the CBP to run jobs on cloud infrastructures assumes that the user possesses proper CBP credentials. In addition to CBP, there are many DCIs that have such requirements (e.g., gLite, ARC, or UNICORE). If gateway providers would like to expose workflows for their users with nodes configured to use such infrastructures, then users will face the difficulty of managing the proper credentials for actually submitting the workflows. To eliminate this need, WS-PGRADE/gUSE has been extended with the robot certificate extension that enables workflow developers to assign predefined credentials for jobs that require some sort of authentication with the DCI to which they are targeted.

Figure 10.8 shows the interface for setting a CBP robot credential for a workflow node set to run on a cloud infrastructure.

After the computing resource has been set, the *Create association...* button has to be clicked, and the CBP (or some other DCI-specific, depending on the target DCI) credentials have to be set in the pop-up window that appears. From this point, it is impossible to modify the target infrastructure and the executable of the workflow node as long as the given association is not removed.

Once a workflow with nodes set to use DCIs requiring authentication but with proper robot certificates assigned to these nodes is exported to the local repository, users not possessing the required credentials will be able to import and actually run the workflow with the robot certificates assigned. The use of robot certificates will be completely hidden from the user; the only thing the user will see is that the workflow can be submitted and is running properly.

10.4.3 Cloud-Based AutoDock Workflows

We are going to present the way to migrate an existing workflow to the cloud based on the AutoDock Vina application.

The migration of the generator and collector nodes was required to follow the following steps:

- Reconfigure the nodes to run on the cloud: For this the node's configuration window (see Figure 10.3) had to be opened; and the node's *type* had to be set to cloudbroker; the *name* to platform; the *resource, region* and *instance type* to MTA SZTAKI.

- Assign robot certificates to the nodes: As accessing the cloud resources requires CBP credentials but it is not feasible to ask every end user of the customized gateway to have a valid CBP credential, a CBP robot credential had to be set for the nodes. A dedicated account for the AutoDock users in the CBP with access granted to the MTA SZTAKI OpenNebula cloud infrastructure has been created, and this account has been set during the nodes' configuration as seen in Figure 10.8.

The migration of the parameter sweep node was a bit more difficult, as this node of the original workflow was set to run on a volunteer DG. In the case of DG applications, the executable resides on the DG server and may consist of multiple files (one "main" executable and supporting files), such as for the Vina DG application. The DG application is only referenced by name in the WS-PGRADE workflow's configuration.

To migrate such an application to the cloud, all executable files of the DG application need to be collected from the DG server and the workflow must be reconfigured to submit these files from the portal server. In the case of a multifile application, there are two options: either the workflow node has to be configured to run the main executable and additional input files

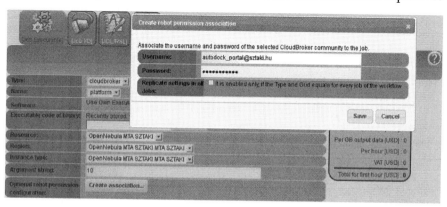

FIGURE 10.8
Robot certificate association.

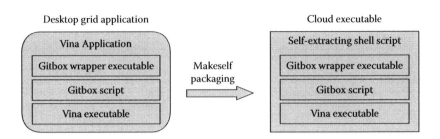

FIGURE 10.9
Multiexecutable desktop grid application bundled into a self-extracting shell script.

representing the supporting files that have to be added to the workflow node or a single self-extracting archive can be created based on all the executable files. Simply, this self-extracting archive should be specified as the application to run for the workflow node.

In the case of the Vina application, we followed the second approach, with the help of the *makeself* UNIX tool [10]. Figure 10.9 illustrates this multiple executable file problem and the solution. The executables belonging to the DG AutoDock Vina application are bundled into a single self-extracting shell script that can be used as the executable for the Vina node in the Vina workflow when configuring the node to run on the cloud.

10.5 AutoDock Gateway

This section explains how to provide user-friendly gateway interface for the end-user scientists who are not interested even in the workflows. They just want to use the AutoDock workflows as black box applications and parameterize and run them in an efficient and fast way in the connected cloud systems. In the previous sections, we showed how to use WS-PGRADE/gUSE to develop and configure the workflows to run in clouds. In this section, we show a unique feature of our gateway technology. This is the end-user view of the WS-PGRADE gateway that can be easily created without any further programming by simply reconfiguring the gateway. In this view, the end-user scientists cannot see the workflows, only their parameter options, through which they can specify their required input parameters. Then, the gateway automatically executes the AutoDock workflows in the connected clouds in an optimal way.

Once the AutoDock workflows are ready, it is simple to create an end-user mode portal for them. Only two steps have to be performed: create templates of the workflows and set the end-user mode as default on the portal.

The template is a special workflow type in WS-PGRADE/gUSE, containing all the properties of a workflow and restrictions on which properties

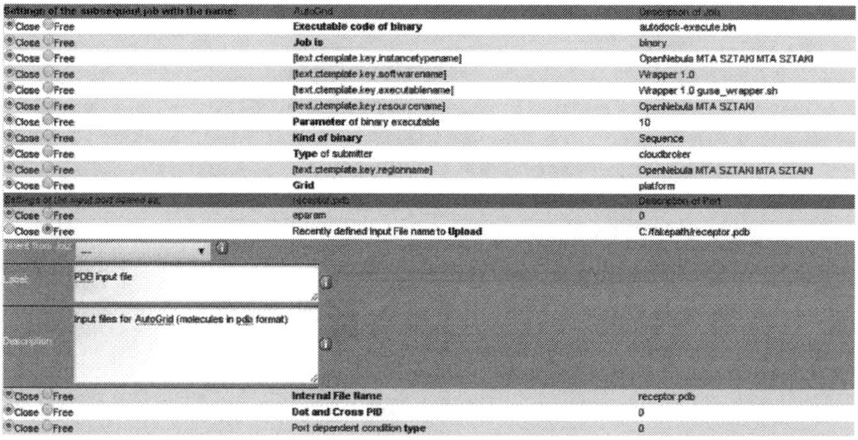

FIGURE 10.10
Template configuration.

can be modified and which cannot. Those properties that can be modified may have a short description, which can be assigned during the template's configuration. In Figure 10.10, the AutoDock workflow's receptor.pdb input file (belonging to the AutoGrid job) is configured: We have set the Upload property to free, meaning that the users will be able to upload their own files to be used as receptor.pdb by the AutoGrid job. Other properties are closed, meaning the users will not be able to modify (or even see) them. Once a template is ready, it can be exported to the internal repository, so that other users of the portal may run them.

Once all the templates are ready and have been exported into the internal repository of WS-PGRADE, the portal can be set into the end-user mode. In this mode, new users will receive only the End User role. The setting can be performed in Liferay's Control Panel, and the process is described in the WS-PGRADE/gUSE Admin Manual [11] in detail. Once this is set, any new user registering to the portal will be an end user and will see only a restricted set of portlets. Figure 10.11 shows the portlets available for end users in the case of the AutoDock portal: Only workflow importing and configuration/execution are possible; workflow creation, storage browsing, and other advanced features are hidden from the end users. Of course, the visibility of the different portlets can be fine-tuned; this process is also described in the WS-PGRADE/gUSE Admin Manual.

Execution of workflows in the end-user mode is really simple: First, the desired workflow has to be imported (select End User/Import as in Figure 10.11 and select the desired workflow as shown in Figure 10.12).

Once the workflow is imported, it can be configured (see End User/Configure in Figure 10.11) by clicking on the "Configure" button in the workflow list.

After the workflow's configuration, the workflow can be executed, and the execution can be monitored. Figures 10.13 to 10.15 show the configuration

FIGURE 10.11
End-user mode on the AutoDock Portal.

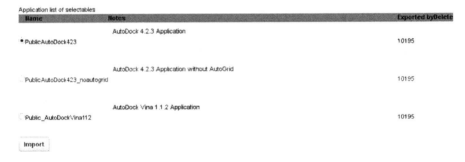

FIGURE 10.12
Select the workflow to import.

FIGURE 10.13
Configuration of a workflow in the end-user mode.

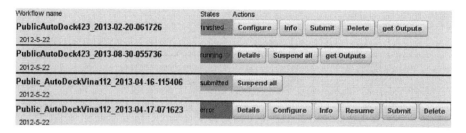

FIGURE 10.14
List of workflows imported.

FIGURE 10.15
Workflow progress monitoring in the end-user mode.

of workflow parameters, the list of workflows imported, and the details of a workflow's execution in the end-user mode, respectively. As can be seen, a progress bar is presented about the workflow's execution, so users can follow their experiments' progress visually.

10.6 Execution Experiences and Performance

This section describes how the AutoDock workflows are actually executed on cloud systems. Performance measurements show the possible optimizations of the gateway back-end mechanism and the CBP to minimize the execution time.

The selected application to perform the performance measurement was the AutoDock Vina workflow, with the input set size of 8,500 ligands. The measurements have been performed by distributing this input set among 25, 100, 250, and 1,000 jobs. Each scenario has been executed on the DG and the cloud. Although we performed the measurements in the different scenarios, we are only presenting our experiences with the 1,000-job scenario as the others showed similar results. The measurements were conducted on the SZTAKI Cloud with 25 virtual machines allocated for the experiments and the EDGeS@home DG with variable and unpredictable active clients (about 2,000 to 3,000 at a time) available.

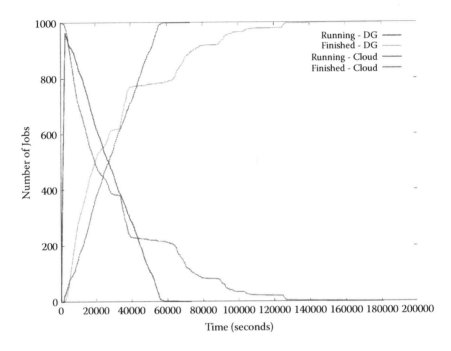

FIGURE 10.16
Processing Vina inputs in 1,000 jobs.

Figure 10.16 shows the case when the inputs were split into 1,000 jobs. The *x* axis of the figure represents the elapsed time, and the *y* axis represents the number of (running and finished) jobs. As can be seen, it took more than twice as long to process the jobs on the DG as on the cloud. In the case of both the DG and the cloud execution, we can see a short "running up" period, followed by a steep processing phase; finally, the last jobs' processing slows.

The job processing figure of the DG case is a bit steeper; thus, processing jobs on this DCI (with 2,000 to 3,000 active clients) is a bit faster than in the cloud (with 25 processors). However, in the case of the volunteer DG, we can clearly observe the tail effect, which means that the last 10%–20% of the jobs required nearly as long an execution time as the first 80%–90% of the jobs. The tail effect is missing in the case of the cloud execution; hence, the overall execution time is much shorter on the cloud. Notice that the tail effect is a well-known problem of volunteer computing, and there are several ways of eliminating it [12, 13]. One of the possible solutions is exactly the support of large volunteer DGs with relatively small dedicated clouds that run the last 10%–20% of the jobs concurrently with the DG. It has been presented [12] that with such a technique the tail effect can be reduced significantly.

10.7 Related Research

This section compares our research with others' targeting similar objectives in distributed computing (such as grid or cloud) environments. This section clearly shows our major contributions to this field.

There are several research projects investigating how biomolecular applications, particularly molecular docking simulations, can be run on distributed computing platforms. Some examples of DCI-based molecular docking simulations are detailed in this section. Most of these experiments are on grid computing resources with rare exceptions currently for the utilization of clouds.

Tantar et al. [14] gave an overview of current efforts on how large-scale parallel computing is applied to molecular simulations. The authors are also involved in the Docking@Grid project [15] that aims to define the optimal deployment architecture for grid-based molecular docking simulations and provide the accurate definition of the molecular energy surface.

The WISDOM project [16] is an international initiative to enable a virtual screening pipeline on a grid infrastructure. WISDOM was the first large-scale in-silico docking experiment on public grid infrastructure. The project has developed its own meta-middleware that utilizes the EGI (European Grid Infrastructure) production infrastructure and is capable of submitting and executing a large number of jobs on EGI resources. Although the WISDOM production environment is capable of submitting any kind of application, the flagship application of the project is AutoDock.

Tantoso et al. [17] described a similar approach for Globus-based grids. A web interface has been developed by the authors to execute experiments using AutoDock3.05 on target grid resources. A small workflow automates the process, which includes the preparation of the receptor, creation of parameter files, calculation of grid energy, and finally the actual docking of the molecules.

Cloud-based molecular docking environments are currently hard to find in the literature. The only example we know about was written by Kiss et al. [18]; the authors described the implementation of similar molecular docking experiments on Windows Azure-based clouds. However, that implementation was closely coupled with the Azure infrastructure, and the user interface is less flexible, making further improvements difficult.

There are also examples for the utilization of higher-level user interfaces for molecular simulations, all based on grid computing infrastructure. The Australian BioGrid portal [19] uses the DOCK [20] molecular docking software for the simulations. This work is part of the Virtual Laboratory Project that aims to utilize grid technologies for solving large-scale compute and data-intensive applications in the area of molecular biology. The project uses the Nimrod Toolkit and the World Wide Grid test bed [21] to conduct the experiments.

The European Chemomentum project developed a collaborative environment based on the UNICORE grid middleware technology to support a wide range of applications in the area of natural and life sciences [22]. Among other applications, the project targeted the AutoDock docking software. The Chemomentum environment also supports the creation and execution of workflows on UNICORE resources.

10.8 Conclusions

As we showed in the section on related research, other AutoDock solutions are tailored to the specific grid or cloud environment. The advantage of our solution comes from its flexibility. First, it is easy to generate various AutoDock gateway workflows for different types of users having different IT expertise and different biological simulation targets. Second, these workflows can be easily reconfigured to run in various DCIs, including clusters, supercomputers, grids, DGs and clouds. In the current chapter, we showed the case when the workflows running originally on DGs were migrated into cloud resources. Even in the cloud environment, it is extremely easy to reconfigure the workflows to run them in various clouds, such as Amazon, IBM, OpenNebula, and OpenStack, due to the integration of WS-PGRADE/gUSE with CBP. This flexibility, of course, can be applied for workflows developed in other fields of science, making the WS-PGRADE/gUSE gateway technology widely usable in many different areas of science and commercial activities as demonstrated by the various science gateways developed in the EU FP7 project SCI-BUS.

Acknowledgment

The research leading to these results has been supported by the European Commission's Seventh Framework Programme (FP7/2007-2013) under grant agreements 283481 (SCI-BUS) and 312579 (ER-Flow).

References

1. Morris, G. M., D. S. Goodsell, et al. Automated docking using a Lamarckian genetic algorithm and an empirical binding free energy function. *Journal of Computational Chemistry* 19(14): 1639–1662, 1998.

2. P. Kacsuk. P-GRADE portal family for grid infrastructures. *Concurrency and Computation: Practice and Experience* 23(3): 235–245, 2011.
3. Eucalyptus. Home page. http://www.eucalyptus.com/.
4. OpenStack. Home page. http://www.openstack.org/.
5. OpenNebula. Home page. http://opennebula.org/.
6. SCI-BUS. Home page. http://www.sci-bus.eu/.
7. Heindl, H., et al. High throughput screening for ligands and inhibitors of carbohydrate modifying enzymes. Proceedings of the 2nd Glyco-Bioinformatics Beilstein-Institut Symposium, June 27–July 1, 2011, Potsdam, Germany. http://www.beilstein-institut.de/glycobioinf2011/Proceedings/Greenwell/Greenwell.pdf.
8. EF-Flow. Home page. http://www.erflow.eu/.
9. CloudBroker Platform. Home page. https://platform.cloudbroker.com/.
10. Makeself. Make self-extractable archives on Unix. 2008. http://megastep.org/makeself/.
11. Gottdank, T. Administrator manual and cookbook. 2013. https://sourceforge.net/projects/guse/files/3.5.8/Documentation/gUSE_Admin_Manual.pdf.
12. Delamare, S., G. Fedak, D. Kondo, and O. Lodygensky. *SpeQuloS: A QoS Service for BoT Applications Using Best Effort Distributed Computing Infrastructures.* Technical report.
13. Pataki, M., and A. C. Marosi. Searching for translated plagiarism with the help of desktop grids. *Journal of Grid Computing* 11(1): 149–166, 2013. doi:10.1007/s10723-012-9224-5. http://dx.doi.org/10.1007/s10723-012-9224-5.
14. Tantar, A.-A., et al. Docking and biomolecular simulations on computer grids: status and trends. *Current Computer-Aided Drug Design* 4(3): 235–249, 2008. doi:http://dx.doi.org/10.2174/157340908785747438.
15. Docking@Grid Project. http://dockinggrid.gforge.inria.fr/index.html (accessed July 1, 2013).
16. Jacq, N., et al. Grid-enabled virtual screening against malaria. *Journal of Grid Computing* 6(1): 29–43, 2008. doi:10.1007/s10723-007-9085-5.
17. Tantoso, E., et al. Molecular docking, an example of grid enabled applications. *New Generation Computing* 22(2): 189–190, 2004. doi:10.1007/BF03040958.
18. Kiss, T., et al. Large scale virtual screening experiments on Windows Azure-based cloud resources. *Concurrency and Computation*, accepted for publication, scheduled for October 2013.
19. Gibbins, H., et al. The Australian BioGrid portal: empowering the molecular docking research community. Proceedings of the 3rd APAC Conference and Exhibition on Advanced Computing, Grid Applications and eResearch, September 2005, Gold Coast, Australia. http://eprints.qut.edu.au/3780/1/3780.pdf.
20. Ewing A. (ed.). DOCK Version 4.0 Reference Manual. San Francisco: University of California at San Francisco (UCSF). 1998. http://www.cmpharm.ucsf.edu/kuntz/dock.html.
21. Buyya, R., et al. The Virtual Laboratory: a toolset to enable distributed molecular modelling for drug design on the World-Wide Grid. *Concurrency and Computation: Practice and Experience* 15(1): 1–25, 2003. doi:10.1002/cpe.704.
22. Chemomentum Project. http://www.chemomentum.org/c9m.

11

SaaS Clouds Supporting Biology and Medicine

Philip Church, Andrzej Goscinski, Adam Wong, and Zahir Tari

CONTENTS

Summary

Cloud computing has started to change the way research in science, in particular biology and medicine, is being carried out. By utilizing different cloud models, biological and medical researchers can take advantage of scalable resources that can be accessed on demand. However, there are also disadvantages in using the cloud, for example, usability issues in infrastructure-as-a-service (IaaS) clouds, limited language support in platform-as-a-service (PaaS) clouds, and lack of specialized services in software-as-a-service (SaaS) clouds. To resolve known issues, we propose the development of research clouds for high-performance computing as a service (HPCaaS) to enable researchers to take the role of cloud service developer. A prototype of our proposed cloud framework has been developed and a case study provided that demonstrates how HPCaaS research clouds can simplify genomic drug discovery via access to cheap, on-demand high performance computing (HPC) facilities. Cloud-based technologies—in all of their many varieties—have completely transformed enterprise information technology (IT). These technologies have revolutionized how users access computational resources, empowering users with on-demand access to computational resources exposed as services, their high scalability, and availability, and providing services through easy-to-use web forms. With the world moving to web-based tools to support business activities such as share trading and e-commerce, as well as daily life activities such as online shopping and banking, it is no wonder that computational biologists and pharmaceutical companies are also moving toward cloud-based e-science (web-based online science) to conduct their research (Subramanian 2012). This chapter presents an SaaS cloud framework to support genomic and medical research. By first investigating how HPC is delivered on clouds, the problems encountered by researchers utilizing the cloud to run HPC applications are identified. To solve these issues, a research cloud framework is proposed that incorporates aspects of currently used e-science and cloud solutions that support research (in biology and medicine). This framework simplifies cloud access and cloud resource management while allowing researchers to take the role of a cloud service developer. A prototype of our proposed cloud framework, called Uncinus, was then implemented and validated through a case study that demonstrates how research clouds can simplify personalized medicine via access to cheap, on-demand HPC facilities.

11.1 Introduction

In recent years, a number of cloud-enabled tools have been developed to support e-science, aiming to support collaboration between scientists, make the use of computer systems easier, and decrease the time for data analysis. Tools developed specifically for computational biology research include scientific workflow systems such as Galaxy (Goecks et al. 2010); web portals for analyzing and sharing genomic data such as expression-package (EXP-PAC) (Church, Goscinski, and Lefèvre 2012); and dedicated sequence-processing software such as Bowtie (Langmead et al. 2009). Running these scientific applications, in many cases, requires a huge amount of computational power to execute complex algorithms or to process big data. High-performance computing (HPC) can provide computer facilities that perform the large and complex simulations and database searches required for research within reasonable time frames. However, using HPC scientific systems and applications is difficult for many scientists who are not computing specialists. It is also a natural expectation of these discipline specialists to be provided with packages/tools that do not require deep knowledge of programming and system management and allow them to use their specialist backgrounds; these packages or tools should be similar to already available easy-to-use software packages.

HPC requires powerful and expensive computational hardware, data storage, advanced middleware, and sophisticated distributed discipline-oriented applications. The process of managing HPC resources requires in-depth system administration skills, for which many scientists are not prepared. Furthermore, due to their high initial purchase price and maintenance costs, HPC resources are only affordable for rich institutions. As a result, these resources are shared by many researchers, which leads to long waiting times for application execution. Thus, many researchers cannot access HPC infrastructures when needed; they often scale down their applications to reduce waiting times. It is these barriers that have hindered many researchers in achieving innovative discoveries for which they must rely on HPC resources.

A response to the problem faced by discipline specialists lies in cloud computing (Goscinski, Brock, and Church 2011). Clouds promise to relieve the pressure put on the demand of affordable, scalable, and on-demand HPC resources that can provide users faster turnaround times on their experiments. Providing users faster turnaround times on their experiments using clouds has been one of the major issues promised to be addressed in a new version of A Grid and Virtualized Environment (AGAVE) (2012). Public cloud vendors, including Amazon's Elastic Compute Cloud (EC2) (Amazon 2010), have provided solutions specifically designed for running HPC applications. EC2 is an excellent example of an infrastructure-as-a-service (IaaS) cloud offering raw processing and storage services. Other vendors provide platform-as-a-service (PaaS) clouds where users can access an integrated software platform for building HPC applications themselves as well as running HPC applications on cloud resources. Examples include Microsoft's Azure (Chappell 2009) and Google's AppEngine (Gibbs 2008). Furthermore, these clouds also provide the ability to scale on demand as the users' requirements change, accelerating the discovery of new knowledge in various fields of research. Clouds can also provide software on demand; examples of software-as-a-service (SaaS) clouds include science cloud, a drug discovery information management technology (Accelrys 2011). Thus, discipline specialists now have access to on-demand, scalable, and pay-as-you-go HPC facilities. However, while clouds alleviate the costs of procuring required information technology (IT) resources, the cost and time of learning how to prepare an HPC cloud and its applications remain a problem to many users.

The rest of this chapter is divided as follows: Section 11.2 investigates how HPC applications are delivered on clouds. The types of HPC applications that suit clouds are examined, as well as how these applications can be provided to researchers.

Section 11.3 presents a framework for publishing cloud resources and cloud applications. This framework incorporates methodologies used by current e-science and research clouds to simplify the development of SaaS applications.

Section 11.4 describes an implementation of the proposed framework. The outcome of this section is Uncinus, a prototype research cloud that supports the publication of software and cloud resources.

Section 11.5 consists of a case study in which Uncinus is used to analyze genomic cancer data. By building a genetic profile of cancer tissues, the cancer subtype was identified, which has ramifications in providing personalized treatment for cancer. Through this case study, it is shown how medical software can be published, exposed, and deployed on cloud resources without the need for complex deployment procedures.

Finally, in Section 11.6 a conclusion of the work carried out and an analysis of achieved results are presented.

11.2 Delivering HPC on Clouds

To deliver HPC on the cloud, an understanding of the cloud, software to be run, and cluster management is required. Using this knowledge, the user must profile the HPC application to be run and select the right cloud resources. This resource selection process has a large effect on the time and cost of running HPC applications in the cloud. Once resources have been selected, the cloud must be configured to enable HPC applications. Most cloud resources that support HPC are provided at the IaaS level (in the form of virtual machines). The user must be able to configure these virtual machines into the form of a cluster, installing middleware and schedulers and deploying HPC applications. The complexity of enabling HPC on the cloud is beyond the scope of most biology and medical researchers. However, solutions exist in the form of e-science applications and research clouds that aim to lessen the computing knowledge required to carry out research on HPC resources and the cloud.

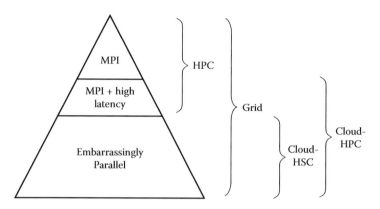

FIGURE 11.1
Workflows and HPC/HSC systems.

11.2.1 Supporting HPC Applications in the Cloud

Traditionally, clouds have been created for business, not to support HPC. However, these days, clouds can support some HPC workloads. Clouds are oriented to support high-scalability computing (HSC) rather than HPC, although with the improvement of communication performance they are becoming a major tool for HPC. Like clusters and grids, clouds also capitalize on distributed resources for applications. A question could be asked regarding what kind of HPC applications could be executed on a cloud. An answer to this question is provided in Figure 11.1 (Mell and Grance 2009). This figure also specifies the scope of our research in terms of workload and HPC/HSC clouds.

Some public cloud vendors, including the Amazon's EC2 (Amazon 2012) have provided solutions specifically designed for running HPC applications. EC2 is an excellent example of an IaaS cloud in that computing infrastructures such as computers, storage devices, and networks are provided to users. There are also private and community IaaS clouds in Australia, including NeCTAR (Kirby 2012), that provide on-demand computing resources to academics and researchers to run HPC applications. Thus, discipline specialists now have on-demand access to HPC facilities that they need with flexible pay-as-you-go pricing methods. HPC clouds give users the opportunity to test and run their parallel applications in the cloud at a price and performance level within what would otherwise be unviable budget constraints. They also provide the ability to scale on demand as the users' requirements change, accelerating the discovery of new knowledge in various fields of research. Thus, clouds can provide discipline specialists faster turnaround times on their experiments.

While the benefits of cloud computing are numerous, delivering HPC on cloud resources is complicated. Currently, before researchers are able to fully utilize HPC clouds, they must understand how clouds are designed. Time must be taken to select cloud resources and enable HPC on the cloud. Selection of the cloud resources has a huge effect on runtime of an HPC application (Church and Goscinski 2011). Non-HPC-enabled clouds are the ideal platform for running embarrassingly parallel applications (e.g., common genetic analysis applications such as mpiBLAST). Embarrassingly parallel applications can take full advantage of cloud scalability to reduce the time and cost of analysis. However, performance studies (Goscinski, Brock, and Church 2011; Expósito et al. 2013) have shown that, when running communication-bound applications, clouds should make use of hypervisors with low overhead and high-speed interconnection.

Once cloud resources have been selected, additional steps must be carried out to enable the use of clouds as a distributed computing system. The steps taken are dependent on the cloud service model. At the IaaS level, this involves construction of a virtual cluster, compilation, and deployment of distributed software. These tasks were previously the job of system administrators and are beyond the scope of most discipline (even computing) researchers. PaaS is aimed at developers, providing users with a development environment and automating the deployment of resources. The problem of this approach is that the user has limited access to development tools and programming languages, thereby limiting the scientific applications that can be deployed. At the SaaS level, the user is able to access HPC applications through graphical interfaces; however, the user is reliant on whichever cloud services have been made available. In specialist research areas such as gene expression profiling and drug discovery, such software would have expensive licenses or not be readily available.

In summary, the complex process of selecting cloud resources and configuring the cloud is beyond the scope of most noncomputing researchers. However, a number of solutions exist to simplify the use of HPC applications and cloud resources.

11.2.2 Solutions Supporting Research on the Cloud

An analysis of the current state of projects and development of computing-based packages and tools to support researchers leads to two major areas: (1) e-science tools based on web application programming interfaces (APIs) and grids (clusters or clusters of clusters) and (2) research clouds, in particular research applications exposed as cloud services (SaaS). Through e-science tools, researchers can run complex software without directly interacting with computing resources; examples include HubZero, P-GRADE, and AGAVE.

HubZero (McLennan and Kennell 2010) is an open-source software platform for creating dynamic websites that support scientific research and educational activities and promote scientific collaboration using primarily the grid infrastructure. By using HubZero, a scientific gateway (website) containing discipline-specific resources, including software applications as well as data repositories, can be formed, and users of the scientific gateway can contribute by putting their own applications and data into the gateway for sharing.

The P-GRADE Grid Portal (Kacsuk 2011) is a web-based, service-rich environ-ment for the development, execution, and monitoring of workflows and workflow-based parameter studies on various grid platforms. P-GRADE Portal hides low-level grid access mechanisms by high-level graphical inter-faces, making even non-grid expert users capable of defining and executing distributed applications on multi-institutional computing infrastructures. Workflows and workflow-based parameter studies defined in the P-GRADE Portal are portable between grid platforms without learning new systems or reengineering program code.

AGAVE (2012) is a software development tool, developed by the Texas Advanced Computing Center (TACC). AGAVE seeks to make the separation of science and computing a bit easier by providing a set of REST APIs for performing distributed and grid computing. AGAVE excels in its ability to provide a holistic view of distributed heterogeneous systems that may span organizational domains into a single, cohesive platform on which modern web applications can be built. The next version of AGAVE promises to include new types of systems, such as public and private clouds, to give users faster turnaround times on their experiments.

While e-science applications are easy to use and thus appealing to research-ers, they are time consuming and require specialized knowledge to develop. Similar to e-science applications, SaaS allows researchers to access complex software with minimal computing knowledge. Development of SaaS is per-formed through research clouds that focus on simplifying access to cloud resources while leaving software development and exposure to a service provider. Examples of research clouds include Aneka, extensions to Globus and Nimrod, and the HPC Hybrid Deakin (H2D) cloud.

Aneka (Calheiros et al. 2012) is a framework for development, deployment, and management of cloud applications. Through a middleware approach, it provides modules to monitor cloud resources. Development in Aneka makes use of predefined programming methods (Task, Threads, MapReduce, and Parameter Sweep), each with different scheduling and execution method-ologies. Aneka relies on a software engineer to develop and expose services.

Recent work from the University of Chicago (Liu et al. 2012) deployed a bioinformatics workflow across local and Amazon EC2 resources. Combining the features of Galaxy and Globus allows for a robust research cloud that sup-ports automated graphical user interface (GUI) generation, software sharing, and workflow deployment. During workflow deployment, data were trans-ferred through a web interface and resources selected manually through creation of a topology file.

Work by Bethwaite et al. (2010) extended the Nimrod tool family to support the Amazon EC2 cloud, allowing access to grid and cloud resources. Four methods of scheduling are available based on user requirements: limited by budget (time), unlimited budget (time optimal), limited by time (cost), and limited by budget and time (none). Nimrod scheduling divides resources into slots based on the processor cores per instance and number of cores required per job.

The last solution is the H2D cloud (Brock and Goscinski 2012), which provides services to discover compute resources and deploy data and applications. This cloud platform is capable of utilizing both local and remote computational services for single large, embarrassingly parallel applications. In this solution, compute resources are published to a dynamic broker service that monitors the state of available compute resources.

By providing scientific software at the SaaS level, it is possible to minimize the computational knowledge requirements needed to access cloud resources. SaaS eliminates the time-consuming tasks of software deployment and hardware setup/management, in particular resource selection and allocation; however, due to the specialized nature of state-of-the-art research, there is limited incentive for attracting cloud software service providers.

11.2.3 Conclusion

While originally their purpose was to support business applications, cloud providers have moved to support HPC applications. These HPC clouds have large amounts of memory, computing power, and high-speed network interconnections. By using these clouds, users can access HPC resources on demand without the need for supporting staff or purchasing expensive hardware. However, to deliver HPC on the clouds, a complicated setup process must be undertaken. Care must be taken to select cloud resources that suit the HPC application being run. Cloud resources must then be configured to allow HPC applications to be run; often, this involves the construction and management of a virtual machine cluster. The computing knowledge required to configure, access, and use cloud resources makes clouds unsuitable for the majority of researchers.

To support research on clouds, access to resources and complex software must be provided to researchers with limited computing background. Two areas that have shown success in bridging the knowledge gap between computing and research are e-science and research clouds. The approach taken by e-science applications and research clouds relies on the abstraction of computing resources from the application logic. While the tools generated from these approaches appeal to researchers, they are not an ideal solution for specialized research. The development of e-science applications requires a multidisciplinary skill set, while the research cloud approach relies on financially motivated providers.

The investment required to develop services for specialized research areas (with a limited market) is not attractive for service providers looking to make a profit. Therefore, the solution is to devise a research cloud that enables researchers to take the role of cloud developer. This research cloud should implement scheduling and execution as well as enhanced features relating to service composition and resource discovery. Such a cloud can incorporate the features found in e-science packages: the application sharing supported by HubZero, P-GRADE's low-level abstraction methodology, and AGAVE's separation of scientific logic from computing. A potential cloud solution was investigated through the development of a research cloud framework.

11.3 A Research Cloud Framework

Known issues can be resolved by developing a unified cloud framework that allows researchers to easily deploy and expose HPC scientific applications in public clouds as services (Wong and Goscinski 2013). Each of these SaaS cloud services is such that it abstracts both the complex deployment effort and the tedious command line execution style of HPC scientific applications into just a web form for scientists to comfortably perform computational research in the clouds.

The basis of our cloud solutions is a framework (see Figure 11.2) that aims to deliver HPC applications to scientists as SaaS cloud services. This framework enables two different processes: cloud software development and cloud service publication. During cloud software development, each HPC application is described by a set of attributes and their associated values. Three major attributes of an application service are (1) a virtual machine image where a targeted HPC application has been properly installed and configured, (2) a web form where parameters for the HPC application would be collected and then passed to the API of the HPC application, and (3) a host location for service invocation, (e.g., SaaS resources). During cloud service publication, these attributes are published to an HPC application

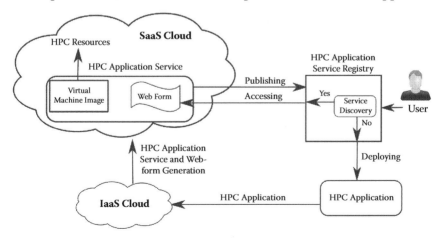

FIGURE 11.2
An overview of HPC cloud framework.

services registry. The HPC application services are proposed to be stored in such a manner that their discovery and selection are easy. This implies that the invocation advice and at least two attributes of an application service, its host location (SaaS cloud) and its web form, must be published. It is proposed to employ a dynamic broker (Brock and Goscinski 2009; Goscinski and Brock 2010) of resources and services to publish application services, thus allowing other end users to learn of the newly deployed application.

11.3.1 Framework Scope

In response to the major problems faced by discipline specialists in using HPC clouds (see Section 11.2), three services are proposed: (1) automation of HPC application deployment, (2) automation of HPC application service and web form generation, and (3) HPC application service registry and formation and application of a wiki-like knowledge base for interface regeneration and HPC application customization. These research areas and the relationships among them form our cloud framework.

11.3.1.1 Automation of HPC Application Deployment

To enable SaaS development by biology and medical researchers, there is a need to automate aspects of the HPC application deployment. Supporting this application deployment process requires at least two levels of abstraction: (1) low-level deployment that consists of methods to install and configure an HPC application in a virtual machine and (2) high-level deployment that consists of methods to save an image of the preconfigured virtual machine and construct an API of the HPC application that together form a deployable unit.

Low-level deployment focuses on automating the installation and configuration of any software application on any computer system. The most common approach taken to automate low-level deployment is seen in package management systems such as the Advanced Packaging Tool (APT) for the Debian GNU/Linux distribution and its variants (Calinou 2012). These tools automate the retrieval, configuration, and installation of software packages, either from binary files or by compiling source code. However, such methods are focused on single machines and are not designed for remote installation on HPC clusters. To enable automated deployment of HPC applications on clouds requires secure access to remote resources and automated resource selection.

High-level deployment focuses on automating virtual machine construction and configuration. Methods encapsulate an HPC application into a virtual machine image and an API of the HPC application; this will form a deployable unit that can be exposed and easily accessed by users as an SaaS cloud service. To support high-level deployment in the cloud, there is a need for tools that can set up an HPC execution environment consisting of software library dependencies, compilers, schedulers, and HPC middleware in a virtual machine environment. Compatibility with low-level deployment will allow for services to be built from the virtual machine level, with software configured and deployed on top of standard virtual machine templates.

11.3.1.2 Automation of HPC Application Service and Web Form Generation

To turn deployed applications and virtual machines into services, they need to be exposed through a graphical interface. The development of interfaces allows for the abstraction of both the application deployment and the command line execution style of HPC cloud applications. To simplify this process, a mechanism is required to automatically transform any HPC application, IaaS cloud execution ready, into an easy-to-use service to be executed in SaaS clouds (Brock and Goscinski 2009).

In the framework, each application is described by a set of attributes and their associated values. The three major attributes of an exposed application service are (1) the location of the virtual machine image where a targeted HPC application has been properly installed and configured, (2) input and output parameters for the HPC application, and (3) service invocation information (e.g., an SaaS cloud service, which was selected by the user and information on how to invoke it). Using the application attributes described, an interface can be derived. Application parameters can be used to form the controls to specify input and collect results. Through these controls, users must be able to upload data and invoke services. Uploading and downloading data make use of the virtual machine location, while invoke services require execution scripts, taken from the service invocation information.

11.3.1.3 Storage of Application Deployment Information

Combining the automation of application deployment and automation of interface generation allows for the construction of HPC application services. To enable sharing of HPC application services between researchers, this framework requires the construction of an HPC application services registry. Each HPC application service is proposed to be published and stored in such a manner that their discovery and selection are easy. This implies that the invocation information and at least two attributes of an application service, its host location (SaaS cloud) and its web form, must be published.

To enable publication of resources and applications, development of a dynamic broker is proposed (Goscinski and Brock 2010; Brock and Goscinski 2012); this will allow other end users to learn of the newly deployed services. Deployed and exposed HPC applications as SaaS cloud services should be easily discoverable and selectable by users. The proposed use of a registry will allow users to discover and select the required services. By storing application information, it is possible to build a repository of common analysis processes and workflows. This repository can have significant impact on time required of service providers and end users, as deployment information is stored and reused. Depending on the type of service published to the registry, different attributes are required. Publication of resources requires the location (DNS, domain name system), and cloud access information (secure shell [SSH] keys, location, and file system layout). Publication of applications requires the input/output details of the application, invocation information, and hardware requirements (operating system [OS], RAM, central processing unit [CPU], network, etc.).

11.3.2 Using the Framework

In the proposed framework, when a user wants to conduct a scientific discovery by executing HPC applications on clouds, the user first contacts the HPC application service registry. The outcome of the service discovery and the user's preference of the HPC application service for the targeted HPC application can lead to two different scenarios for the user.

In the first scenario, particularly in the case of a discipline researcher who does not have programming and system administration skills, the HPC application service of the user's interest is found. On selection of the cloud service, resources are selected, and the application deployment service sets up and configures the cloud. While this is happening, the automated interface generation service constructs a user-friendly discipline-specific interface for the requested HPC application service. Access to the cloud service is conducted through the generated interface.

In the second scenario, the HPC application service of the user's interest is not found. The user, who has programming and system administration skills, would have to deploy a new targeted HPC application in an IaaS cloud. The proposed automatic HPC application deployment system can automate parts of this process. The outcome of this process would be either a virtual machine image that contains a copy of the properly installed and configured HPC application or a software service (consisting of input/output, invocation information, and hardware requirements) that can be deployed on a virtual machine. At this stage, the cloud service published in the HPC application service registry is readily accessible in an IaaS cloud. The new cloud service generated by the automatic HPC application deployment system is stored for future use in the HPC application service registry. In the next stage, the user can employ the automatic HPC application service and web form generation system to automate the formation of an HPC application service exposing the corresponding HPC application. The HPC application service is abstracted by a virtual machine image and a user-friendly discipline-specific interface that is published in the HPC application service registry and could allow the user to easily access the targeted HPC application in an SaaS cloud.

11.4 Research Cloud Prototype

A prototype of the proposed cloud framework was developed by integrating three components: (1) Amazon EC2 (public IaaS cloud), which provides HPC infrastructure; (2) an HPC service software library (Church, Wong, Brock, and Goscinski 2012) for accessing HPC resources from an IaaS cloud; and (3) an application broker as a web-based platform (Uncinus) for accessing and exposing HPC applications.

11.4.1 Prototype Overview

The overview of the prototype design demonstrating the relationships among the Amazon EC2 service, the HPC software library, and the application broker, is shown in Figure 11.3. Also shown in Figure 11.3 is our view of the cloud service stack where different cloud services would be found. At the bottom (IaaS) layer, the Amazon EC2 was used to provide cloud infrastructure services. Supported HPC applications are installed in virtual machines, and their images are saved and stored in Amazon EC2.

In the middle (high-performance computing as a service, HPCaaS) layer, an HPC software library was developed to expose and access Amazon EC2 services. The software library provides users a higher level of HPC services, such as constructing and managing computer clusters. A web form exposes the operations of the HPC software library, allowing the user to start cloud jobs (selecting the type and amount of resources that are required) and

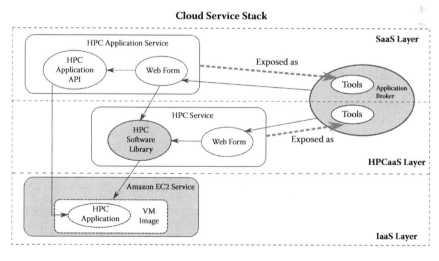

FIGURE 11.3
Implementation overview of HPC cloud framework.

terminate resources when jobs are complete. This HPC software library interface is made available through the application broker.

On the top (SaaS) layer, a service for supporting HPC application deployment was developed as follows: First, an API of the HPC application was constructed. This API acts as a program stub for its corresponding HPC application, which, when deployed, is installed in a virtual machine on the Amazon EC2 cloud. Second, a web form exposing the HPC application service is generated. This HPC application service can then be published to the applications broker to be exposed as a software service. It should be noted that each HPC application service would access the HPC services in the HPCaaS layer and the HPC application installed and stored in a VM image at the bottom IaaS layer through its web form.

The components that make up the framework are described in more detail in the following sections. Section 11.4.2 presents the operation of Amazon EC2, which provides cloud infrastructure services to the research cloud prototype. Section 11.4.3 describes the construction of the HPCaaS model and how it is able to abstract Amazon EC2 resources. Section 11.4.4 describes the construction of a research cloud called Uncinus; this cloud provides an application broker to deploy and expose applications. Uncinus also exposes the services provided by the IaaS and HPCaaS layers through easy-to-use web interfaces.

11.4.2 Amazon EC2: The Public IaaS Cloud Service Provider

Amazon EC2 provides various computer instance types specifically designed for running HPC applications. Our work has utilized the elastic compute cloud services and the elastic block store services to deploy and run HPC applications. The simplest way to use the EC2 services is by accessing the Management Console (Amazon Web Services [AWS] 2013). After logging on to the AWS Management Console, a user can carry out HPC activities with EC2 by performing the tasks of

1. Selecting the desired Amazon Machine Image (AMI) and launching computer instances;
2. Installing and configuring software;
3. Establishing connection to the computer instances, running applications, and handling data transfer;
4. Terminating computer instances and releasing resources.

This approach of accessing a public HPC cloud service is quite ad hoc and could be tedious for discipline scientists who have little background in HPC. On top of the work in launching, connecting, and terminating AWS computer instances, discipline scientists are also forced to deal with many details to set up and configure an HPC cluster and install middleware and software applications before the system is available for any actual scientific investigation. For this reason, we have developed the HPCaaS model and implemented its software libraries, which provide high-level functions for obtaining HPC resource from IaaS clouds.

11.4.3 The HPCaaS Model: Providing an HPC Service

We have proposed an enhanced cloud service model by including the HPCaaS layer (see Figure 11.3), to abstract HPC resources, including both the hardware (networks, storage, and servers [physical/virtual]) and the software (operating systems, middleware, and user-level HPC applications). This implies that these HPC resources can be exposed to the cloud community as service software libraries. These software libraries are (1) a set of virtual machine images with prebuilt HPC applications and (2) a set of program scripts that can be used to create, manage, and terminate a computer cluster. They support accessing HPC applications and transferring data between users and the computer cluster.

11.4.3.1 Providing an HPC Service Library for Amazon EC2

For each of the IaaS cloud service providers, the corresponding HPCaaS services are grouped together and packaged as a service library. In this case, the HPC service software library for Amazon EC2 is implemented. Currently, we have provided its implementation on the Linux platform.

11.4.3.2 Management and Application Access Scripts

To implement the cluster management and HPC-application accessing scripts of the HPC service software library, the EC2 command line tools (Kay 2012) are used. The only dependencies are Perl and CURL, which are normally included in many Linux distributions. Due to the use of the EC2 command line tools, the HPCaaS model is compatible with Amazon-like clouds that support the same API. Examples of such clouds include private clouds OpenStack (OpenStack Project 2012), Eucalyptus (Nurmi et al. 2009), and research cloud NeCTAR (Kirby 2012).

Through the command line tools, we have implemented a set of bash shell scripts to handle features such as HPC cluster creation and termination, HPC-application job submission, and retrieval of results of HPC-application execution (Wong and Goscinski 2012). A list of selected bash scripts is shown in Table 11.1.

11.4.3.3 AMIs with Prebuilt HPC Applications

When accessing any HPC application in the proposed cloud framework, an HPC cluster must be created and started on EC2, where two pieces of information are required: (1) an AMI for creating instance(s) and (2) the

TABLE 11.1

Selected Management and Application Access Scripts

Shell Script	Description
awsConnect [-n number]	Connect to Amazon EC2 and start *n* number of Cluster Compute instants (an EC2 cluster)
awsReady	Check readiness of the EC2 cluster created for accepting instructions
awsTerminate	Terminate the EC2 cluster created
awsRun	Run an HPC application in an EC2 cluster
awsTransfer	Transfer file(s) to an EC2 cluster
awsCheck	Check readiness of the result
awsCollect	Collect result files from Amazon to a local host for postprocessing

total number of instances for the HPC cluster. The former contains all information necessary to boot instances of software, such as operating system, middleware, and a specific HPC application. The latter quantifies the size of the HPC cluster, which in turn defines the maximum computing power to be provided.

Although there is a large collection of public AMIs available from Amazon and the EC2 community, it is the responsibility of an HPCaaS service provider to provide a customized AMI for each HPC application it supports. In our example, all AMIs provided in the HPCaaS service library for Amazon EC2 run the Ubuntu Linux operating system, the OpenMPI middleware, and Linux-based open-source HPC applications. These AMIs have been made publicly accessible. Users can access them by referencing the AMI IDs. Consequently, the construction of an HPC cluster and the installation and configuration of an HPC application has been abstracted into selecting an AMI from Amazon to use.

11.4.4 Uncinus: An Application Broker and HPCaaS Cloud Solution

Uncinus integrates the HPCaaS libraries into a web platform, allowing researchers access to Amazon EC2 resources configured for HPC. To implement the features described in the framework, a number of services are provided. These services fall into the SaaS and HPCaaS layer previously described (see Figure 11.4). The SaaS layer provides application broker functionality through the Application to Interface Parser (AIP) and Resource Deployment Recorder (RDR); these services write data to the mySQL database management system (DBMS). The HPCaaS layer provides the functionality to configure cloud resources for HPC and deploy applications. Cloud resources configured for HPC are provided by the Cloud Resource Allocation (CRA) service, while application deployment is supported through the Secure Data Transfer (SDT) service. The CRA service incorporates the access scripts developed as part of the HPCaaS model (see Section 11.4.3.2) and as such interacts with Amazon EC2 using the Amazon API.

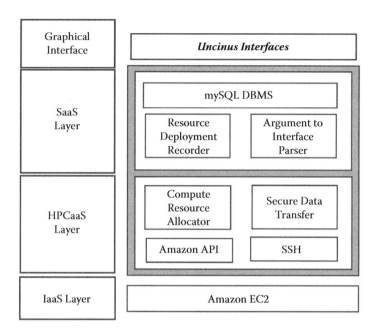

FIGURE 11.4
Uncinus software overview.

The application broker allows an application service provider to publish AMIs or applications. To publish an AMI, a machine image identifier, username, and working directories need to be provided. When publishing applications, the machine image identifier in which to install the package, installation instructions, any required installation files, a list of input arguments, and output controls need to be provided. The RDR stores these data in a mySQL database on behalf of the cloud service provider. The AIP service translates the input arguments recorded by the broker into equivalent web controls, allowing for dynamic interface generation. End users accessing the Uncinus system can select from the published AMIs and application modules to customize their cloud deployment.

The HPCaaS model communicates with Amazon EC2 to provide cloud resources to the user. When a user starts a cloud job, the CRA module creates the necessary private keys and security groups before requesting resources from Amazon EC2. Once the CRA can successfully access the virtual machine, the SDT module is used to deploy application modules to the virtual machine instance.

11.4.4.1 Features of Uncinus

Uncinus combines the HPCaaS model with an application broker to provide the features of the framework detailed in Section 11.3. Automated application deployment is provided by the HPCasS layer through the CRA and SDT services. Automated interface generation is provided by the SaaS layer through the AIP service. Application deployment storage is also provided through the SaaS layer using the RDR service. The automated application deployment services are implemented through use of the HPCaaS model and the RDR service. When deploying cloud resources, the cluster management and HPC-application accessing scripts developed as part of the HPCaaS model are utilized. Uncinus exposes these scripts through a graphical interface, allowing for users to select the number and type of cloud resources required. When deploying application resources, the HPCaaS model configures cloud resources and the RDR service carries out software deployment. Supporting application deployment is an attribute resource selection method. This resource selection method uses published application requirements to identify the optimal cloud resources to carry out the requested software service.

Automated interface generation is provided through the AIP service, which translates published deployment data into web interfaces. For an interface to be automatically generated, each input and output argument must be given a type. To support the typing of program arguments, the AIP service is designed around an XML-like language. Through this language, a user can define a number of common input and output types, including

- <upload>—Secure transfer of data files; the service provider can specify the file name.
- <text>—A text-based argument, which is substituted into the execution script when the service is invoked.
- <config>—Exposes a configuration file through a text control, allowing for direct manipulation of services. This type of control is often used to expose and configure tools, such as databases and web servers.
- <webpage>—Exposes existing web interfaces; used when deploying web server applications on the cloud.

When a service is invoked through Uncinus, input arguments and outputs are passed to the AIP service. The resulting web form contains controls, created based on the typing information.

Application deployment storage is implemented by linking the RDR service with a relational database, resulting in the deployment of an application broker. This broker supports publication of not only applications but also cloud resources as services. The publication process differs depending on the type of service that is deployed. Publication of applications as a service requires installation information, execution information, and application requirements to be specified. Publication of cloud resources is simpler than software, requiring only the cloud location and deployment information. Users publish and access applications as services through a series of web interfaces, the appearance and operation of which are described in Section 11.4.4.2.

11.4.4.2 Accessing Uncinus

Access to Uncinus features are provided through easy-to-use web interfaces. Through these interfaces, users can publish software and cloud resources as a service and access published services through the broker, which incorporates automated application deployment.

The interface used to publish applications is seen in Figure 11.5. Using this interface, a service provider begins by assigning a descriptive name to the application service (Application Name); this name is used during application discovery. For each application, the broker stores the installation procedures (Install Script) and running procedures (Running Script) to be undertaken on remote compute resource(s). Files required by the application service are also stored by the broker (Files); stored files could be source code, binaries, or application data. Stored files can be accessed during the installation and execution procedures. Service providers must provide the broker with information on how to invoke the application service (AppLocation). Information about how input parameters (Arguments) and expected program output (Results) are displayed is stored as XML. Optionally, additional usage information about the application (Manual) can be published. Finally, each application

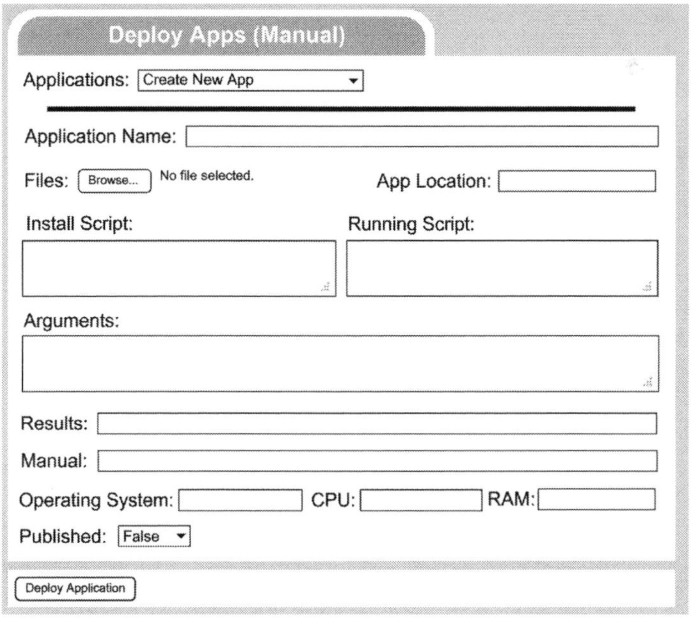

FIGURE 11.5
Uncinus application publication interface.

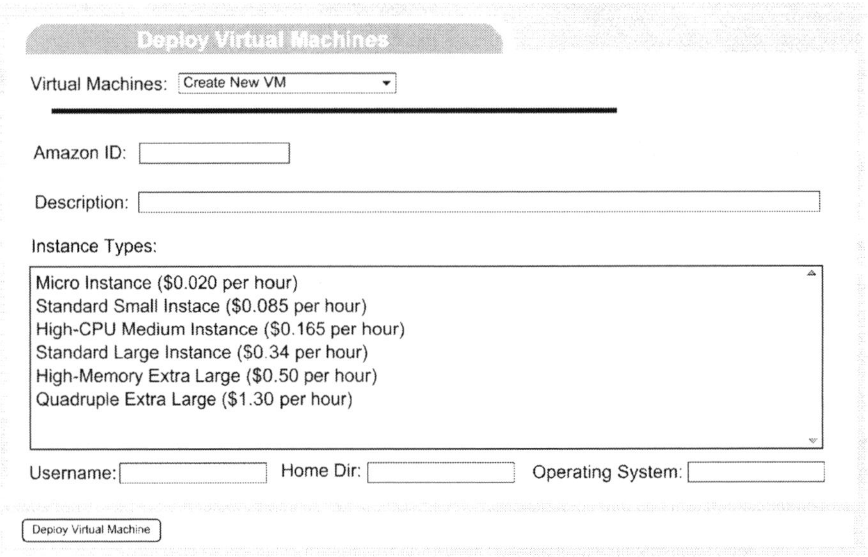

FIGURE 11.6
Uncinus virtual machine image publication interface.

service must define hardware requirements and the amount of resources utilized during execution. The broker stores the required operating system (Operating System); CPU utilization, consisting of number of nodes, cores, and clock speed (CPU); and memory utilization in gigabytes (RAM). On successful publication, the service can be flagged for public viewing (Published).

The interface used to publish virtual machines is shown in Figure 11.6. Using the provided interface, the service provider specifies the location of the virtual machine instance using an identifier given by the cloud provider (Amazon ID). The instance types that each virtual machine supports are also provided (Instance Types); from the instance type, cost and the hardware specification of each virtual machine can be determined.

The interface used to invoke services is shown in Figure 11.7. Through this interface, a user is able to give a descriptive name to a job for identification purposes. Users then select from published resources to create a pool of hardware; any number of published virtual machines can be added to the pool. Optionally, users can also select from a list of public (and their private) software services (Application Modules). Once cloud resources and application modules are selected, the service can be invoked by clicking the "Submit Job" button. During the deployment process, the selected pool of resources is allocated and made ready for HPC execution. If a software service has been selected, resources are selected and configured for the application. The user is then directed to the graphical interface generated using the automated interface generation parser, exposing the service controls.

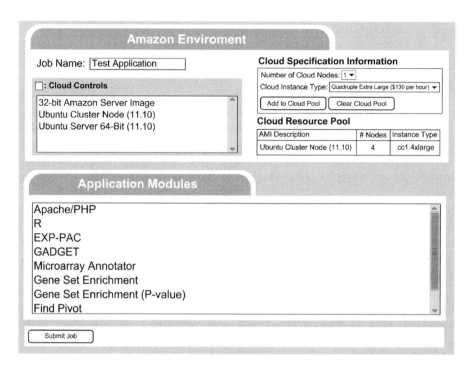

FIGURE 11.7
Uncinus service selection interface.

11.5 Case Studies

Using the prototype, a case study was carried out that demonstrates how cloud platforms can simplify genomic drug discovery via access to cheap, on-demand HPC facilities. An EXP-PAC (Church, Goscinski, and Lefèvre 2012) image was created, deployed, and exposed through our cloud framework prototype. Using EXP-PAC, genetic cancer tumor data is analyzed and annotated. These annotated data are used to create a gene expression profile. The generated gene expression profile is an initial step in the identification of the cancer subtype and possible treatment methods (Beltran and Rubin 2013).

11.5.1 Cloud Deployment

To deploy software as services using the HPCaaS model described, cloud images must be created and then exposed through Uncinus. Cloud images are derived from preexisting templates and stored by the cloud provider (in this case, Amazon EC2). Each image is given a unique identifier by Amazon, which is then published to Uncinus. On invoking the published virtual machine service, Uncinus communicates with the Amazon cloud, requesting the stored image and the required number and type of resources. This deployment procedure was carried out for EXP-PAC.

To set up the EXP-PAC cloud image, a complex deployment procedure is carried out (see Figure 11.8). First, an Ubuntu server AMI is selected from the Amazon EC2 web interface and launched. Second, as this image is not from a trusted source, steps must be taken to ensure the image has not been compromised. Antivirus scans are performed, and the Ubuntu image is updated to ensure there are no vulnerabilities. Next, using the Ubuntu software repository, LAMP is installed; this software package contains the principal components (APACHE, PHP, and mySQL) to build a viable general-purpose web server. PHP and APACHE are configured, increasing the POST and upload data limit to support large data upload and analysis. EXP-PAC is then placed into the web server directory and configured to use the mySQL database. To enable the HPC features of EXP-PAC, openMPI and bioconductor are also deployed on this server. The Amazon cloud image is then stored in its modified form for future use.

Publication of the EXP-PAC virtual machine image to Uncinus was performed through a web interface (see Figure 11.9). The virtual machine publication interface allows users to specify information about the published cloud image that is used during deployment. The attributes required to publish a virtual machine image are the AMI ID of the cloud image, a description of the published cloud image, the supported instance types of the image, log-in information, the home directory, and the OS utilized by the cloud image.

11.5.2 Workflow Execution

Once software has been deployed on the cloud, users can execute exposed applications through published interfaces. To utilize the HPC normalization methods provided by EXP-PAC, this case study was run on four cluster compute instances (64-bit, dual-quad core; 23 GB RAM).

Breast cancer tumor RNAseq data (GSM721140) was downloaded from the National Center for Biotechnology Information (NCBI). These data contained 44.8 million sequence fragments, which were mapped (aligned) to the human reference genome. To be analyzed, a number of preprocessing steps were carried out on the data. First, SAMtools (Li et al. 2009) was used to convert the downloaded data to a human-readable format. The converted data were imported into HTSeq (Anders 2010) (run in union mode, nonstranded), by which sequence fragments that matched known genes were sorted and counted. The output of HTSeq was a list of genes and the amount of times they appeared in the tumor.

In addition to the list of expressed genes, it was necessary to identify the amount of mutations that had occurred in each gene. A mutation score was given to each sequence by counting the bases that differed from the reference genome. This process resulted in the creation of two data sets, a count of present

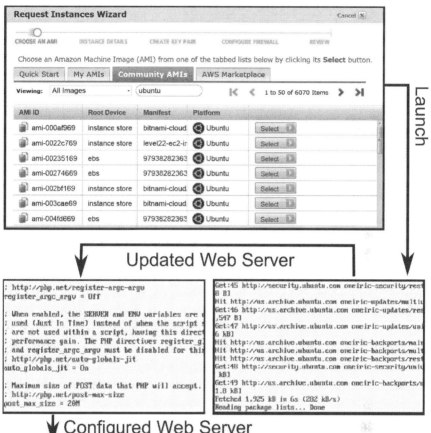

© 2008 - 2013, Amazon Web Services, Inc. or its affiliates. All rights reserved. Privacy Policy Terms of Use

FIGURE 11.8
EXP-PAC Amazon machine image setup.

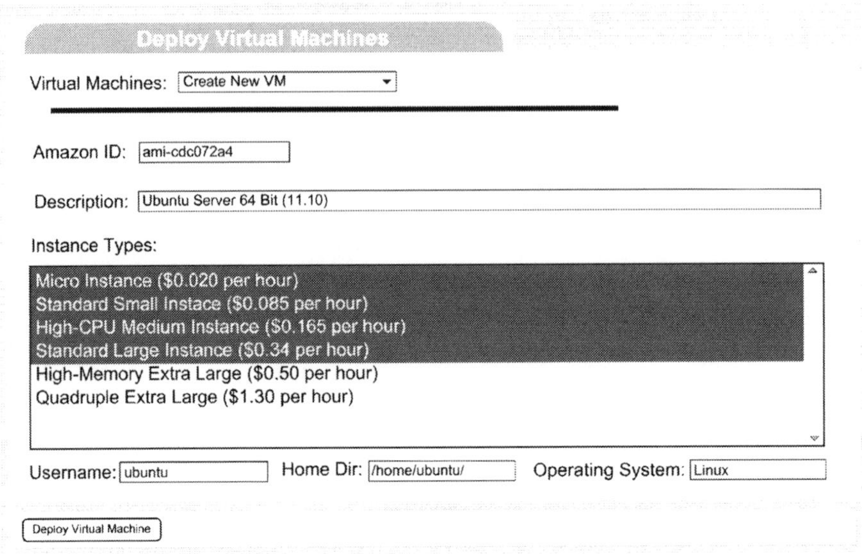

FIGURE 11.9
EXP-PAC publication process.

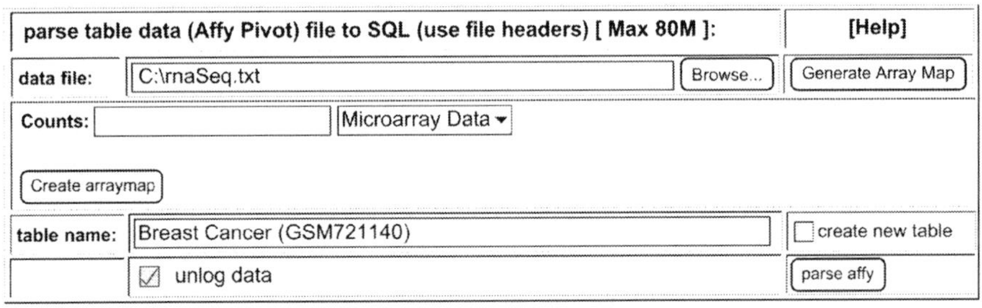

FIGURE 11.10
EXP-PAC data upload interface.

genes and a file containing mutation scores for each gene. These data sets were normalized and uploaded into EXP-PAC using web interfaces (see Figure 11.10).

Using the work of the Cancer Genome Atlas Network (2012), a list of genes was defined for each subtype of cancer: luminal A, luminal B, basal-like, and human epidermal growth factor receptor 2 enriched (HER2E) (see Table 11.2). The luminal A and B signatures overlap, both involving the mutations of tumor protein 53 (TP53), Phosphatidylinositol 3-kinase (PIK3CA), and mitogen-activated protein kinase kinase kinase 1 (MAP3K1). However, luminal A can be identified through the mutation of GATA binding factor 3 (GATA3) and Forkhead Box protein (FOXA1), which are unique to this cancer subtype. Basal-like tumors have high levels of mutation in the TP53, retinoblastoma 1 (RB1), and breast cancer 1 early onset (BRACA1) genes. HER2E

TABLE 11.2

List of Mutated Genes Indicative of Each
Breast Cancer Subtype

Luminal A	Luminal B	Basal-like	HER2E
TP53	TP53	TP53	TP53
PIK3CA	PIK3CA	RB1	PIK3CA
MAP3K1	MAP3K1	BRCA1	PIK3R1
GATA3			PTEN
FOXA1			

TABLE 11.3

List of Highly Expressed Genes Indicative of
Each Breast Cancer Subtype

Luminal A/B	Basal-like	HER2E
ESR1	PIK3CA	FGFR4
XBP1	KRAS	EGFR
MYB	EGFR	HER2
RB1	FGFR1	GRB7
	FGFR2	GATA3
	KIT	BCL2
	MET	ESR1
	PDGFRA	

differs from other subtypes by having a high level of PIK3CA mutations and lower frequency of phosphatase and tensin homolog (PTEN) mutations.

As there are significant overlaps in the gene signatures of these cancer subtypes, to improve the accuracy of analysis, nonmutated genes can be utilized. The genes in Table 11.3 are known to be expressed in a nonmutated form in cancer tumors. By utilizing the list of mutated and highly expressed genes, it is possible to improve the accuracy of diagnosis.

The mutated gene list (see Table 11.2) and highly expressed gene list (see Table 11.3) were loaded into EXP-PAC. Using these lists, queries were performed on the breast cancer data set (see Figure 11.11). For each breast cancer subtype, mutated genes were identified by performing a keyword search for genes in the mutated gene list and looking for results with high mutation scores. Highly expressed genes were identified through a keyword search of gene symbols, this time looking for genes that appeared more than once. To ensure that displayed genes were present in the tumor data, the intensity filter was set to return genes with an intensity greater than 0.

Results (see Table 11.4) showed that, out of the genes known to undergo mutation during breast cancer, only TP53 was expressed in a mutated form. Mutated genes common to other subtypes were not present, which is indicative of a basal-like breast cancer tumor. Examining the expression of

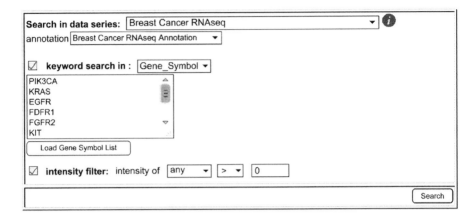

FIGURE 11.11
EXP-PAC query interface.

TABLE 11.4

Mutated Genes Present in the
GSM721140 Data Set

Gene Symbol	Counts	Mutation Score
TP53	4	3
PIK3CA	0	0
MAP3K1	0	0
GATA3	0	0
FOXA1	0	0
PTEN	0	0

TABLE 11.5

Basal-like Indicative Genes Present
in the GSM721140 Data Set

Gene Symbol	Counts	Mutation Score
KRAS	7	2
EGRF	2	0
FGFR2	1	0
MET	1	1

nonmutated genes known to be present in basal-like breast cancer tumors (see Table 11.5) further validated this finding. Four of eight known basal-like indicative genes were shown to be present in the GSM721140 data set. Two genes, V-Ki-ras2 Kirsten rat sarcoma viral oncogene homolog (KRAS) and epidermal growth factor receptor (EGRF), were shown to be highly regulated, while fibroblast growth factor receptor 2 (FGFR2) and mesenchymal-epithelial transition factor (MET) were present.

11.6 Conclusion

Cloud computing can benefit biology and medical studies by supporting collaboration with other scientists and providing cheap access to large amounts of HPC resources. However, utilizing these HPC cloud resources requires users to undertake a complex setup procedure that consists of cloud resource selection, configuring cloud resources for HPC, and application deployment. Researchers looking to take advantage of cloud computing to carry out HPC analysis require an understanding of cloud architecture, software to be run, and cluster management, which is beyond the scope of most researchers. While e-science and research cloud solutions simplify access and execution of applications on HPC resources, they do not solve the difficulties in developing and exposing analysis tools. As such, researchers still require enough computing knowledge to utilize and manage large amounts of HPC cloud computing resources, or they become reliant on financially motivated cloud service providers. In response to this problem, the following question was asked: How can a researcher with limited computing knowledge become a cloud service provider?

An SaaS cloud framework was developed with the aim to simplify the procedures undertaken by service providers, in particular during service deployment and exposure. By identifying and automating common procedures, the time and knowledge required to develop cloud services is minimized. Three procedures were identified and became the focus of automation: application deployment, interface generation, and service storage. By automating application deployment, the computing knowledge required by biology and medical researchers to access cloud software is reduced. By automatically deriving an interface from the inputs and outputs of a service, the programming requirements to expose software as a service are reduced. Finally, by providing service storage, the time taken for analysis is reduced (through the reuse of deployment information).

Implementation of the SaaS cloud framework was realized in the form of Uncinus, a research cloud prototype compatible with Amazon EC2. Fulfilling the requirements of automated application deployment required that cloud resources be configured for HPC. To support this functionality, a new cloud model called HPC as a service (HPCaaS) was proposed that automatically configures cloud resources for HPC. To support automatic interface generation, an XML-based language was developed, as was a parser to translate inputs and outputs to web interfaces. To fulfill service storage requirements, an application broker was built for clouds that supported publication of cloud resources and software services. By implementing features of the framework, Uncinus benefits biological and medical researchers by simplifying the process of developing and deploying software on cloud resources configured for HPC. Cloud services can be built by publishing attributes (input/output, computational requirements, etc.) through easy-to-use web interfaces.

Using Uncinus, a case study was carried out that utilized personalized genomics analysis to perform diagnosis of a patient's breast cancer tumor to identify targeted drug treatment strategies. During this study, genomic analysis tools were installed onto a cloud resource. This cloud resource was then published, deployed, and exposed though Uncinus with minimal user interaction. This case study clearly demonstrated how the automated procedures (proposed by the framework) allow biology and medical researchers to access and deploy cloud services. Through the cloud, researchers take advantage of flexible pricing and on-demand resources that can provide faster turnaround times on their experiments. In the case of embarrassingly parallel applications, such as the presented personal genomics case study, clouds can fully utilize scalability to analyze thousands of cancer genomes at once (something not possible on HPC clusters).

A research cloud solution, like the one presented and implemented in this chapter, allows biology and medical researchers to apply the power of the cloud to their research. This solution was developed by abstracting Amazon EC2 resources from application logic and identifying and automating common methods used in service deployment. In this way, SaaS clouds can be developed that simplify the process of using HPC cloud computing for research.

References

Accelrys. 2014. Accelrys Science Cloud, https://www.sciencecloud.com.

AGAVE. 2012, December. http://sourceforge.net/projects/agaveapi.

Amazon. 2010. *Amazon Elastic Compute Cloud: Getting Started Guide*. Edited by AWS. Amazon, www.amazon.com/dp/B007Q4H6KK.

Amazon. 2012. Amazon EC2 instance types. http://aws.amazon.com/ec2/instance-types/.

Amazon Web Services. 2013. AWS Management Console—Amazon Web Services 2013. http://aws.amazon.com/console/.

Anders, Simon. 2010. HTSeq: analysing high-throughput sequencing data with Python. http://www-huber.embl.de/users/anders/HTSeq/doc/overview.html.

Beltran, Himisha, and Mark A. Rubin. 2013. New strategies in prostate cancer: translating genomics into the clinic. *Clinical Cancer Research* 19(3):517–523.

Bethwaite, Blair, David Abramson, Fabian Bohnert, Slavisa Garic, Colin Enticott, and Tom Peachey. 2010. Mixing grids and clouds: high-throughput science using the Nimrod Tool family. In *Cloud Computing*, edited by N. Antonopoulos and L. Gillam. London: Springer, 219–237.

Brock, M., and A. Goscinski. 2009. Attributed publication and selection for web service-based distributed systems. Paper presented at 2009 World Conference on Services-I, July 6–10, 2009, Los Angeles, CA.

Brock, M., and A. Goscinski. 2012. Execution of compute intensive applications on hybrid clouds (case study with mpiBLAST). Paper presented at 2012 Sixth International Conference on Complex, Intelligent and Software Intensive Systems (CISIS), July 4–6, 2012, Palermo, Italy.

Calheiros, Rodrigo N., Christian Vecchiola, Dileban Karunamoorthy, and Rajkumar Buyya. 2012. The Aneka platform and QoS-driven resource provisioning for elastic applications on hybrid Clouds. *Future Generation Computer Systems* 28(6):861–870.

Calinou. 2012. Apt-Debian wiki. https://wiki.debian.org/Apt.

Cancer Genome Atlas Network. 2012. Comprehensive molecular portraits of human breast tumours. *Nature* 490(7418):61–70.

Chappell, David. 2009. *Introducing Windows Azure.* San Francisco: David Chappell and Associates.

Church, P. C., and A. Goscinski. 2011. IaaS clouds vs. clusters for HPC: a performance study. Paper presented at the Second International Conference on Cloud Computing, GRIDs, and Virtualization, at Rome, Italy.

Church, Philip C., Andrzej Goscinski, and Christophe Lefèvre. 2012. EXP-PAC: providing comparative analysis and storage of next generation gene expression data. *Genomics* 100(1):8–13.

Church, P., A. Wong, M. Brock, and A. Goscinski. 2012. Toward exposing and accessing HPC applications in a SaaS cloud. Paper presented at 2012 IEEE 19th International Conference on Web Services (ICWS), June 24–29, 2012, Honolulu, HI.

Expósito, Roberto R., Guillermo L. Taboada, Sabela Ramos, Juan Touriño, and Ramón Doallo. 2013. Performance analysis of HPC applications in the cloud. *Future Generation Computer Systems* 29(1):218–229.

Gibbs, Kevin. 2008, April. Google App Engine Campfire One transcript. http://code.google.com/appengine/articles/cf1-text.html.

Goecks, Jeremy, Anton Nekrutenko, James Taylor, and the Galaxy Team. 2010. Galaxy: a comprehensive approach for supporting accessible, reproducible, and transparent computational research in the life sciences. *Genome Biology* 11(8):R86.

Goscinski, Andrzej, and Michael Brock. 2010. Toward dynamic and attribute based publication, discovery and selection for cloud computing. *Future Generation Computer Systems* 26(7):947–970.

Goscinski, Andrzej, Michael Brock, and Philip Church. 2011, June. High performance computing clouds. In *Cloud Computing: Methodology, System, and Applications,* edited by L. Wang, R. Ranjan, J. Chen, and B. Benatallah. Boca Raton, FL: CRC Press, Taylor & Francis Group.

Kacsuk, Peter. 2011. P-GRADE portal family for grid infrastructures. *Concurrency and Computation: Practice and Experience* 23(3):235–245.

Kay, Timothy. 2012. Simple command-line access to Amazon EC2. http://aws.amazon.com/developertools/739 (accessed March 2012).

Kirby, Judd. 2012. NeCTAR—Australian Research Cloud. 2012. http://www.nectar.org.au/.

Langmead, Ben, Michael Schatz, Jimmy Lin, Mihai Pop, and Steven Salzberg. 2009. Searching for SNPs with cloud computing. *Genome Biology* 10(11):R134.

Li, H., B. Handsaker, A. Wysoker, T. Fennell, J. Ruan, N. Homer, G. Marth, G. Abecasis, and R. Durbin. 2009. The Sequence Alignment/Map format and SAMtools. *Bioinformatics* 25(16):2078–2079.

Liu, B., B. Sotomayor, R. Madduri, K. Chard, and I. Foster. 2012. Deploying bioinformatics workflows on clouds with Galaxy and Globus Provision. In *The Third International Workshop on Data Intensive Computing in the Clouds*, Salt Lake City, UT.

McLennan, M., and R. Kennell. 2010. HUBzero: a platform for dissemination and collaboration in computational science and engineering. *Computing in Science and Engineering* 12(2):48–53.

Mell, Peter, and Tim Grance. 2009. The NIST definition of cloud computing. *National Institute of Standards and Technology* 53(6):50.

Nurmi, Daniel, Rich Wolski, Chris Grzegorczyk, Graziano Obertelli, Sunil Soman, Lamia Youseff, and Dmitrii Zagorodnov. 2009. The Eucalyptus open-source cloud-computing system. In *Proceedings of the 2009 9th IEEE/ACM International Symposium on Cluster Computing and the Grid*. New York: IEEE Computer Society, 124–131.

OpenStack Project. 2012. OpenStack—Open source software for building private and public clouds. http://www.openstack.org/.

Subramanian, Bhuvaneashwar. 2012. The disruptive influence of cloud computing and its implications for adoption in the pharmaceutical and life sciences industry. *Journal of Medical Marketing: Device, Diagnostic and Pharmaceutical Marketing* 12(3):192–203.

Wong, Adam K. L., and Andrzej M. Goscinski. 2012. A VMD plugin for NAMD simulations on Amazon EC2. *Procedia Computer Science* 9(0):136–145.

Wong, Adam K. L., and Andrzej M. Goscinski. 2013. A unified framework for the deployment, exposure and access of HPC applications as services in clouds. *Future Generation Computer Systems* 29(6):1333–1344.

12

Energy-Aware Policies in Ubiquitous Computing Facilities

Marina Zapater, Patricia Arroba, José Luis Ayala Rodrigo, Katzalin Olcoz Herrero, and José Manuel Moya Fernandez

CONTENTS

Summary

Next-generation e-science applications such as the ones found in smart cities, e-health, or ambient intelligence, require constantly increasing high computational demands to capture, process, aggregate, and analyze data and offer services to users. Research has traditionally paid much attention to the energy consumption of the sensor deployments that support this kind of application. However, computing facilities are the ones presenting a higher economic and environmental impact due to their very high power consumption. In this chapter, we provide a vision of the increasing energy problem in computing facilities with a focus on cloud computing, under the new computational paradigms, and propose solutions from a global, multilayer perspective, describing a novel system architecture, power models, and optimization algorithms. This chapter is organized as follows: Section 12.1 introduces the topic; Section 12.2 briefly describes the related work. Section 12.3 describes a novel system architecture for the global energy optimization of next-generation e-science applications. Section 12.4 describes the power models developed for the architecture, and Sections 12.5 and 12.6 briefly describe some optimization techniques. Finally, Section 12.7 summarizes the most important aspects.

12.1 Introduction

Data centers are easily found in every sector of the worldwide economy. They provide the required infrastructure for the execution of a wide range of applications and services, including social and business networking, web mail, web search, electronic banking, Internet marketing, distributed storage, high-performance computing (HPC), and so on. The increasing demand for higher computer resources has recently facilitated the rapid proliferation and growth of data center facilities. In recent years, population-monitoring applications (such as e-health applications or ambient intelligence), e-science, and applications for smart cities have experienced significant development, mainly because of the advances in the miniaturization of processors and the proliferation of embedded systems in many different objects and applications (e.g., communications, industrial, automotive, defense, and health care environments). Next-generation systems consist of a large set of nodes, distributed among the population. Data obtained by these sensor nodes are communicated to the embedded processing elements by means of wireless connections. Huge sets of data must be processed, stored, and analyzed. To deal efficiently with such computationally intensive tasks, the use of cloud services is devised since cloud computing is emerging as the dominant computer platform for scalable online services.

Thus, the wireless body sensor networks (WBSNs) will be connected not only at the node level but also through a personal digital assistant (PDA) or smartphone to the cloud. Part of the data processing and storage will be local to the node, while another part will be communicated and processed in the cloud, depending on the application, on the state of the batteries, and on security or privacy requirements of the information. This computing environment where the mobile client utilizes mobile network services to communicate with the cloud through the Internet is usually known as mobile cloud computing (MCC) [1].

Recent research has focused on developing energy efficiency policies at the data center level. Some policies have been detected but not successfully proposed as they lack consideration of the global power consumption of the system. They do not take into account that the agents involved in the problem are heterogeneous. Therefore, the energy cost of performing part of the processing in any of the different abstraction layers, from the node to the data center, should be evaluated.

Our proposal develops global energy optimization policies that start from the design of the architecture of the system, with a deeper focus on data center infrastructures, and take into account the energy relationship between the different abstraction layers, leveraging the benefits of heterogeneity and application awareness.

12.2 Related Work

For decades, data centers have only focused on performance, defined as speed. Examples include the TOP500 list of the world's fastest supercomputers (http://www.top500.org), which calculates speed as floating-point operations per second (FLOPS), and the annual Gordon Bell Awards for Performance and Price/Performance at the Supercomputing Conference (http://www.supercomp.org). However, raw speed has increased tremendously over the past decade without relative and proportional energy efficiency. In 2007, although there had been a 10,000-fold increase in speed since 1992, performance per watt was only improved 300-fold and performance per square foot only 65-fold [2].

This huge performance improvement is mainly due to increases in three different dimensions: the number of transistors per processor, the operating frequency of each processor, and the number of processors per system. Collectively, these factors yield an exponential increase in power consumption of data centers that is not sustainable. The focus on just speed has let other evaluation metrics go unchecked. Data centers consume a huge amount of electrical power and generate a tremendous amount of heat. To support these technologies, during 2008 world power consumption exceeded US$30 billion [3] when an average data center consumed as much energy as 25,000 households [4]. About 15% of these costs are due to removing the heat generated throughout the infrastructure [5]. The situation is critical since the numbers are growing. In 2010, the worldwide data center consumption reached 1.5% of global energy, having increased by 56% since 2005 [6].

To this end, major players in the data center and high-end computing markets often negotiate energy deals with electricity suppliers to build or upgrade power substations, near or immediately next to, their computing facilities. Alternatively, when not enough power infrastructures can be built at or near computing facilities, many companies move their computing facilities to the power source (e.g., Google [7], and Microsoft [8]).

In addition to the economic impact of excessive energy consumption, the environmental impact has affected the data center community. The heat and the carbon footprint emanating from cooling systems are dramatically harming the environment. According to Mullins [9], US data centers use about 59 billion kWh of electricity, exceeding US$4.1 billion and generating 864 million metric tons of CO_2 emissions released into the atmosphere.

Both research and industry have recently proposed several approaches to tackle the power consumption issue in data center facilities. Industry has begun to shift the goal from performance to energy, reporting not only FLOPS but also FLOPS per watt and measuring the average power consumption when executing the LINPACK (HPL) benchmark [10]. Today, metrics such as being in the Green500 list [2] are beginning to be of importance. Also, reference companies around the world, such as Google, IBM, or Amazon, are implementing measures to make their data centers more efficient and beginning to measure the power usage effectiveness (PUE) of their facilities.

PUE is one of the most representative metrics and consists of the facility's total power consumption divided by the computational power. PUE close to 1 means the data center is using most of the power for the computing infrastructure instead of it being lost or devoted to cooling devices. Average PUE for 2011 was around 1.83%, which does not represent a sufficient reduction for sustainable infrastructures. According to Amazon data center estimations [11], expenses related to operational costs of the servers reach 53% of the budget, while energy costs add up to 42%, which are broken down into cooling (19%) and power consumption of the infrastructure (23%). Therefore, the cooling problem needs to be approached to restrain the upward trend [3] and to prevent these technologies reaching beyond the limit of sustainability.

Researchers have done a massive amount of work to address these issues and provide energy-aware computing environments. From the data room perspective, previous work addressed the power consumption problem by means of optimizing cooling costs at the resource manager level by assigning longer tasks to servers with lower inlet temperature [5]. From the information technology (IT) perspective, research has proposed solutions to reduce the computational power of servers by means of energy-efficient scheduling techniques [12, 13], resource allocation, and workload assignment mechanisms [14, 15]. For cloud computing applications, virtualization technology has provided a promising way to manage application performance by dynamically reallocating resources to virtual machines (VMs). Several management algorithms have been proposed to control the application performance for virtualized servers [16] and to solve the VM-server mapping problem for power savings [17].

In our proposal, scheduling and resource allocation take into account the global energy consumption, which includes the cooling consumption of the data center and the consumption of the rest of the system: node and PDA or smartphone. So, it is possible to take advantage of the heterogeneity of the system and download only some part of the computation to the data center, while the rest is performed in the PDA.

12.3 Proposed Novel Paradigm

12.3.1 Devised Computer Paradigm

Next-generation applications are usually composed of a large number of sensors, wirelessly connected to the cloud through a mobile processing device. Data centers provide cloud-based data services that can closely match the demand of processing capacity, according to data size and complexity of the analysis algorithms. By sharing data center resources for multiple applications, it is possible to reduce the need for resources, maintaining high utilization rates and reducing energy requirements. To provide adequate energy management, this heterogeneous distributed computing system is tightly coupled with an energy analysis and optimization system, which continuously adapts the amount of processing that is performed in the different layers of the distributed system and the resources assigned to each task.

12.3.2 Energy Optimization System

Figure 12.1 shows the proposed system architecture for the energy optimization of cloud computing in e-science applications. Detailed functions of constituents in the system are summarized as follows:

- Application support network: Applications require a heterogeneous network comprising sensor nodes, data centers, and some kind of interconnection network to drive data from sensors to data centers. Each element has different computation capacity, functional requirements, power consumption characteristics, and so on.

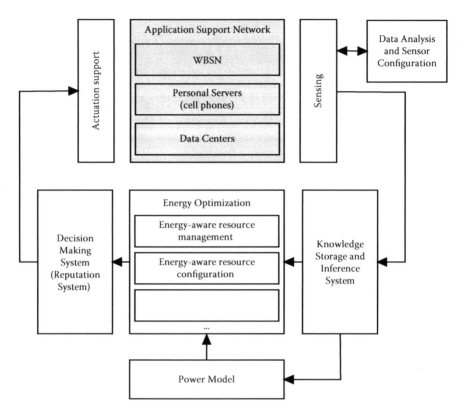

FIGURE 12.1
Overview of the proposed energy analysis and optimization system.

- Sensing infrastructure: Global energy optimization requires a clear understanding of the current state of the network, the characteristics of the different resources, and of the analysis to be performed. Therefore, additional hardware (HW) or software (SW) sensors should be added to the system to obtain insight.

- Data analysis and sensor configuration: Not every sensor has the same importance to understand the power consumption characteristics of the different components. After a careful analysis, the sensing infrastructure has to be configured to provide only the relevant data at the required rate for the power model to be useful and to minimize the energy overhead.

- Storage and inference system: The data provided by the sensing infrastructure has to be stored and statistically analyzed in search of recurrent behaviors that could lead to simple but accurate enough power models to be used for proactive optimizations. Although the data provided by the sensors is low level, simple inference techniques can be used to raise the level of abstraction, for example, to understand the energy demand characteristics of the different analysis applications or the power consumption characteristics of different resources.

- Power model: Complex power models are not adequate for online optimization, as different alternatives should be quickly evaluated against the power model to proactively configure the whole system for minimum energy consumption. These power models can be trained with actual data from sensors to improve the quality and to adapt to variations in the heterogeneous application support network.

- Optimization: Based on the current state of the system, the historic data, and the energy characteristics of application and resources, many optimization algorithms can be executed to enhance one or more aspects of the population-monitoring system. Heterogeneity can be analyzed to always assign tasks to the most adequate resources; resources not being used can be turned off; cooling energy can be taken into account when assigning tasks to resources; and, at the same time, when a group of nodes is detected to behave anomalously, they can be discarded to provide some kind of self-healing mechanism.

- Decision-making system: There are so many aspects that can be optimized (at different levels of abstraction and in different scopes), that it would not be feasible to consider all of them in a single optimization algorithm. Many partial optimization algorithms may propose actions in the network; some of them could even be incompatible with other decisions. We propose the use of a reputation system [18] to compose the decisions provided by multiple optimization algorithms and to adapt to changes in the system by changing the weight of different optimizations.

- Actuation support: Finally, decisions should be executed. Software agents in all levels of the application support network are in charge of reconfiguring their behavior whenever an optimization decision is made.

12.4 Energy and Power Models

To apply energy optimization techniques at all levels, but most importantly to the cloud computing facility, we need to develop power and energy models of the resources of the data center that can be applied to predict the energy consumption of the workload to be executed. In this section, we describe the most important contributors to the energy consumption in data centers, and we present some of the most relevant energy- and power-modeling techniques.

12.4.1 Overall Power and Energy Consumption Breakdown

The main contributors to the energy consumption in a data center are the computing power (also known as IT power), which is the power drawn by servers to execute a certain workload, and the cooling power needed to keep the servers within a certain temperature range that ensures safe operation. Together, both factors account for more than 85% of the total power consumption of the data center, with the other 15% the power consumption due to lightning, generators, UPS (uninterruptible power supply) systems, and PDUs (power distribution units) [6].

$$P_{DC} = P_{IT} + P_{cooling} + P_{others}$$

The IT power is dominated by the power consumption of the enterprise servers in the data center. The power consumption of an enterprise server can be further divided into three different contributors: (1) the dynamic or active power, (2) the static or leakage power, and (3) the cooling power due to the server fans:

$$P_{server} = P_{static} + P_{dynamic} + P_{fan}$$

Dynamic power is the power due to the switching of the transistors in electronic devices; that is, it is the power used to perform calculations. Leakage power is the unwanted result of subthreshold current in the transistors and does not contribute to the microcontroller function. Fan power is becoming a more important contributor by the day to overall server power [19].

Cooling power is one of the major contributors to the overall data center power budget, consuming over 30% of the overall electricity bill in typical data centers [20].

12.4.2 Computing (IT) Power Modeling

12.4.2.1 Static Power Consumption: Leakage Power Modeling

Dynamic consumption has historically dominated the power budget. But, when the integration technology scales below the 100-nm boundary, static consumption becomes much more significant, being around 30%–50% [21] of the total power under nominal conditions. This issue is intensified by the influence of temperature on the leakage current behavior. There are various leakage sources in devices, such as gate leakage or junction leakage, but at present, subthreshold leakage is the most important contribution in modern designs. Therefore, it is important to consider the strong impact of static power as well as its temperature dependence and the additional effects influencing their performance. The current consumption of an MOS device due to leakage current is the one shown in the following equation:

$$I_{leak} = I_s \cdot e^{\frac{V_{GS}-V_{th}}{nkT/q}} \cdot \left(1 - e^{\frac{V_{ds}}{kT/q}}\right) where I_s = 2 \cdot n \cdot \mu \cdot C_{ox} \cdot \frac{W}{L} \frac{kT^2}{q}$$

When $V_{DS} > 100mV$, the contribution of the second exponential is negligible [22], so the previous formula can be rewritten as follows:

$$I_{leak} = I_s \cdot e^{\frac{V_{GS}-V_{th}}{nkT/q}} = B \cdot T^2 \cdot e^{\frac{V_{GS}-V_{th}}{nkT/q}}$$

where technology-dependent parameters can be grouped together in a constant B.

Based on the leakage current equation, we describe the leakage power for a particular machine m as the next equation:

$$P_{leak,m} = I_{leak,m} \cdot V_{DD,m} = B \cdot T^2 \cdot e^{\frac{V_{GS}-V_{th}}{nkT/q}} \cdot V_{DD,m}$$

As can be seen, leakage has a strong dependence on temperature. Even though power models have traditionally disregarded leakage, recent studies are beginning to take it into account. Some cloud computing solutions, such as those in Reference 23, have considered the dependence of power consumption on temperature due to fan speed as well as the induced leakage current. Moreover, taking into account the leakage-cooling trade-offs at the server level by finding an optimum point between the fan power and the leakage power has proven to yield up to 10% energy savings at the server level [24].

In the case of cloud computing, it is especially interesting to take into account the temperature of the different computing resources. The pool of resources that builds the entire cloud infrastructure allows the utilization of those resources most appropriate to the operating situation. Thus, depending on the type of application and the thermal state of the machine, an efficient allocation can be performed that minimizes the static consumption of the computing infrastructure by keeping the unused resources in a low-power state.

12.4.2.2 Dynamic Power Modeling

Dynamic power consumption varies depending on the characteristics of the particular workload to be executed, as well as on the platform where the workload is executed. The same workload can present different energy behavior depending on the target platform, as shown in Figure 12.2, obtained from Reference 25.

To understand and take advantage of these differences, dynamic power has to be modeled. Dynamic power modeling of enterprise servers has recently been tackled via the use of performance counters [26, 27]. Performance

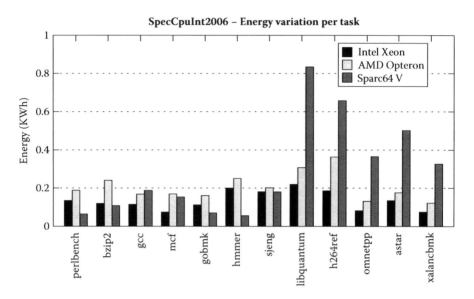

FIGURE 12.2
Energy consumption for SPEC CPU 2006 executed in various servers.

counters are a set of special-purpose registers built into modern central processing units (CPUs) to store the counts of hardware-related events. Because they are integrated into the architecture, polling these counters has a negligible overhead in the performance of the workload being profiled. Modern servers come with a high number of performance counters that can be polled. By collecting performance counters together with information on the power consumption of the server, power consumption can be modeled and thus predicted. Servers are also shipped with a large amount of sensors to collect temperature, fan speed, or power consumption data. These data can be gathered via the Intelligent Platform Management Interface (IPMI) tool (http://ipmitool.sourceforge.net) with negligible overhead. Information of the performance counters can be correlated with power and then regressed to obtain a model for dynamic energy. The performance counters that influence the model vary depending on the system architecture and allow an explanation for the differences in power consumption of the same workload in different servers.

12.4.3 Data Center Cooling Power and Data Room Modeling Techniques

In a typical air-cooled data center room, servers are mounted in racks, arranged in alternating cold/hot aisles, with the server inlets facing cold air and the outlets creating hot aisles. The computer room air conditioning (CRAC) units pump cold air into the data room and extract the generated heat (see Figure 12.3).

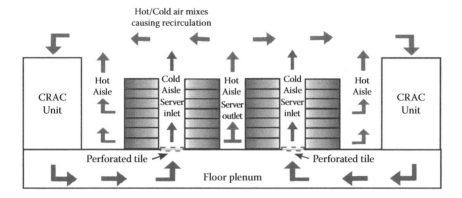

FIGURE 12.3
Diagram of an air-cooled data center room.

The efficiency of this cycle is generally measured by the coefficient of performance (COP). The COP is a dimensionless value defined as the ratio between the cooling energy produced by the air-conditioning units (i.e., the amount of heat removed) and the energy consumed by the cooling units (i.e., the amount of work to remove that heat).

$$COP = \frac{Output\ Cooling\ Energy}{Input\ Electrical\ Energy}$$

Higher values of the COP indicate a higher efficiency. The maximum theoretical COP for an air-conditioning system is described by Carnot's theorem as in the next equation:

$$COP_{MAX} = \frac{T_C}{T_H - T_C}$$

where T_C is the cold temperature (i.e., the temperature of the indoor space to be cooled), and T_H is the hot temperature (i.e., the outdoor temperature; both temperatures in celsius). As the room temperature and the heat exhaust temperature increase, approaching the outdoor temperature, the COP increases and the cooling efficiency improves. According to this, one of the techniques to reduce the cooling power is to increase the COP by increasing the data room temperature.

However, as we increase room temperature, CPU temperature increases and so does leakage power. Therefore, there is a trade-off between the reduction in cooling power and the increase in server leakage power. Previous approaches [29] showed how two different working regions can be found depending on the impact of ambient temperature in leakage power and thus in the total power consumption of enterprise servers. For the lower range of ambient temperatures, the impact of the temperature-dependent leakage is negligible, whereas for a higher-temperature range leakage needs to be considered.

To ensure the reliability of the IT equipment, CPU temperatures should not increase above a certain threshold. The ASHRAE (American Society of Heating, Refrigerating, and Air-Conditioning Engineers) [29] organization publishes metrics on the maximum inlet air temperature for a server, the redline temperature, as well as the appropriate temperature and humidity conditions of the data room environment to ensure that reliability is not affected.

Data room modeling is still an open issue, as the only feasible ways to model the thermal behavior of the data room and be able to predict the inlet temperature of the servers is either by deploying temperature sensors in the data room that take measurements or by performing time-consuming and expensive computational fluid dynamics (CFD) simulations. CFD simulations use numerical methods to analyze the data room and model its behavior. However, these simulations do not often match the real environments and must be rerun every time the data center topology changes.

12.5 Ubiquitous Green Allocation Algorithms

Resource management is a well-known concept in the data center world and is used to allocate in a spatiotemporal way the workload to be executed in the data center, optimizing a particular goal. Traditionally, these techniques have focused on maximizing performance by assigning tasks to computational resources in the most efficient way. However, the increasing energy demand of data center facilities has shifted the optimization goals toward maximizing energy efficiency. Works proposing allocation algorithms have traditionally applied greedy algorithms [30], Markov chain algorithms [31], mixed-integer linear programming (MILP), or mixed-integer nonlinear programming (MINLP) [32] to generate the best task allocation. Most of these approaches do not propose a precise objective function or accurate mathematical formulation of the optimization problem. Although some of these solutions behave well in homogeneous data-center-level scenarios, they do not consider the heterogeneity inherent in smart environment applications. Moreover, MILP solutions do not scale well for larger scenarios with a high number of servers and large workloads to allocate.

Only very recently industry and research started to agree on the importance of environmental room monitoring [33] to improve energy efficiency. Other research [34] presented the data center as a distributed cyberphysical system (CPS) in which both computational and physical parameters can be measured with the goal of minimizing energy consumption from a

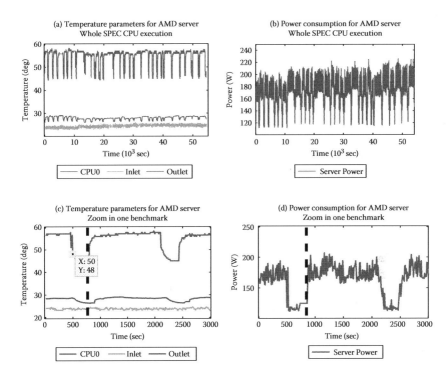

FIGURE 12.4
Temperature and power values for AMD server under SPEC CPU 2006 workload.

jointly computational and cooling perspective. However, these works do not generally apply their solutions in a real scenario.

Our proposal considers not only the heterogeneity that comes from the use of different servers inside a data center facility but also the use of the heterogeneous elements that compose the MCC scenario outside the facility. We leverage the use of nonoptimal lightweight distributed allocation algorithms based on the use of satisfiability modulo theory (SMT) formulas outside the facility. We combine this allocation with MILP-based problems in the data center facility and envision the use of genetic algorithms (GAs) to solve larger resource management problems. We apply these algorithms to real data collected from a completely monitored data room, obtaining inlet and outlet server temperature values, CPU temperatures, server fan speed, server power consumption, and cooling power. Figure 12.4 shows the temperature and power traces obtained from an AMD Sunfire V20Z server when executing tasks of the SPEC CPU 2006 benchmark [35].

12.5.1 SMT Solvers

An SMT solver decides the satisfiability of complex formulas in theories such as arithmetic and uninterpreted functions with equality. An SMT solver is a tool that allows checking whether a certain formula satisfies a condition. SMT solvers are fast and lightweight and thus can be used in nodes with limited resources in a distributed way. Our proposal leverages the idea developed in Reference 36 and proposes that each node of the network, to decide whether to execute a task or offload it to the data center, should run the SMT solver. The SMT solver calculates which tasks of the workload satisfy the conditions to be executed at the node and the amount of tasks that can be executed.

12.5.2 Mixed-Integer Linear Programming

Regarding IT power only, the proposed resource allocation algorithms aim to minimize the overall energy consumption of the data center by assigning tasks in a spatiotemporal way to the most appropriate processors. Mathematically, let us denote by M a set of machines, by P a set of processors, and by T a set of tasks that must be executed. Each processor p belongs to one machine m, denoted as p_m. Each machine m consumes an idle power of π_m. Every task t has a duration and consumes a certain amount of energy over idle depending on the target processor, σ_{tp} and e_{tp}, respectively. The problem consists of finding the most appropriate allocation of tasks t in processors p to minimize the energy consumption, as expressed in the next equation:

$$Min\left\{ \sum_{t\in T, p\in P} k_{tp} \cdot e_{tp} + \sum_{m\in M} \pi_m \cdot \tau^{max} \right\}$$

where k_{tp} is a binary variable that is set to 1 if the task t is executed in processor p. τ^{max} is the time instant at which all the tasks have been executed. As can be seen, the first part of the formula accounts for the dynamic energy consumption, whereas the second part accounts for static power consumption of the servers.

The optimization is subjected to the following constraints:

$$\sum_{p\in P} k_{tp} = 1, \quad \sum_{t\in T} k_{tp} \cdot \sigma_{tp_m} + \gamma_{pm} \le \tau^{max}$$

The factor γ_{pm} is a time offset that represents the amount of time that a processor is occupied (executing previous tasks) when the new job set arrives. In this way, the system can take into account the initial use of processors.

12.5.3 Genetic Algorithms

The previous MILP solution is valid for a data center room with a limited amount of computational resources and an optimization objective that can be expressed as a linear problem. However, when scaling in the number of resources and tasks to allocate, GAs behave much better in terms of performance.

One of the benefits of using a GA is the possibility of tackling a large set of constraints (the maximum temperature of the servers, the available CPU capacity, the required instructions per task, etc.). In this way, the GA defines a vector of n decision variables, a vector of m objectives function, a number of constraints not satisfied, the total energy, and the feasible region in the decision space. The algorithm allows unfeasible solutions, but only when no other alternatives are found.

For the chromosome encoding, each gene represents a decision variable. Because many decision variables are integers, the chromosome uses integer encoding. Thus, some decision variables (like the CPU capacity) are scaled to the integer interval and transformed to a percentage when used in the multi-objective function for evaluation. The evolutionary solver starts with a random population of chromosomes. After that the algorithm involves the population applying (1) the non-dominated sorting genetic algorithm (NSGA-II) standard tournament operator, (2) a single-point crossover operator with probability of 0.9, (3) an integer flip mutation operator, and (4) the multiobjective evaluation. Steps 1 to 4 are applied for a variable number of iterations or generations.

Using this approach, it is possible to obtain optimal energy savings, realistic with the current technology, in much shorter time than traditional algorithms and targeting much more complex environments.

12.6 Resource Selection and Configuration

Cloud computing presents a compelling opportunity to reduce data center power bills. The economic advantages of shifting to a cloud infrastructure are enormous, and current challenges in cloud adoption will be overcome soon, leading a major shift to cloud computing. In this computational context, the goal of techniques like "resource selection" and "configuration" is to offer new services more efficiently by properly selecting and configuring the available resources. The algorithms described in the previous section can be jointly applied with the cloud-specific techniques proposed in this section—virtualization, consolidation, and managing the operating server set—to substantially increase energy savings.

12.6.1 Virtualization

Virtualization allows the management of the data center as a pool of resources, providing live migration and dynamic load balancing, as well as the fast incorporation of new resources and power consumption savings. In addition, a single node can accommodate simultaneously various VMs (based on different operating system environments) that can be dynamically started and stopped according to the system workload and that share physical resources.

Some research work has tried to address the VM provisioning challenge by predicting the workload profile with neural networks and using heuristics to assign the workload [37]. However, to obtain the most energy-efficient setup, the MILP and GA previously described can be used to dynamically assign VMs to physical servers, also deciding the amount of VMs needed to execute a certain workload.

12.6.2 Consolidation

Historically, data centers have been oversized, using a small fraction of their computing resources. Consolidation uses virtualization to share resources and reduces energy consumption by increasing resource utilization. This technique allows multiple instances of operating systems to run concurrently on a single physical node, avoiding wasted physical resources. Consolidation allows reducing the number of operating servers to process the same workload, minimizing the static consumption, which leads us to operating server set and turn-off policies.

Workload allocation algorithms should also take into account the possibility of consolidation. As the number of decision variables and the design space grow larger, GA-based solutions become more suitable for the purpose of efficient VM assignment and consolidation.

12.6.3 Operating Server Set and Turn-off Policies

This technique consists of modifying the active server set by switching off idle hosts when occupancy decreases. Another advantage of cloud computing is that in many applications, such as data mining and web searching, using MapReduce provides outsourcing of the workload. MapReduce, popularized by Google [38], is widely used in application-level energy-aware strategies due to simplified data processing for massive data sets to increase data center productivity [39]. When a MapReduce application is submitted, it is separated into multiple Map and Reduce operations so its allocation may influence task performance. This factor allows leveraging server resources by distributing the workload to achieve the optimal minimum consumption.

One of the issues to consider when implementing this type of policy is the characterization of the use of the data center by customers. The demand for resources reaching the data center is variable and usually follows seasonal patterns depending on the time of day or certain periods of the year. In addition, the data center must be prepared to support peak demand.

Also, the quality of service (QoS) contracted by customers must be satisfied in matters of availability and both execution and response time constraints.

Moreover, the cost of the machine turned on or off to suit the operational farm-to-user demand also must be taken into account. This cost involves two important factors:

- Energy: Consumption of machines when turned off and on again is significant [40]. The energy saved during the period in which servers are switched off should be compensated by this offset energetic cost.
- Delay: The server turn on takes a certain time, so the incoming demand and its variations have to be anticipated. Backup physical machines should be available to host peak requirements.

Currently, one common technique is to apply low-power modes to inactive servers to save static energy [41]. This policy helps minimize delays when activating new machines under peak demand, reducing consumption of idle servers. Many servers offer *sleep* or *hibernate* states, such as *standby* modes, that consume less than active modes with different setup times. Finally, it is necessary to take into account these additional costs in resource configuration policies to minimize energy globally.

This technique can be combined with dynamic voltage and frequency scaling (DVFS). Dynamic consumption can also be reduced by acting on the low-power modes of the machines at runtime, but only if this policy does not violate QoS requirements contracted by users. Modifying the frequency, voltage, or both varies the response time, affecting the completion of services and applications. Decreasing the frequency or operating voltage reduces dynamic power consumption during the execution of a workload. Also, during idle periods, the static consumption is minimized at low voltages and frequencies.

Therefore, if QoS restrictions are not strict, energy savings in the computing part can be increased by the efficient application of the presented techniques.

12.7 Conclusions

Cloud computing, MCC, or even modern HPC start with data centers. While we can dream of a world in which anyone is allowed to sell their excess computing capacity as virtualized resources to anyone else or where the ubiquitous sensing of information is processed by a center kilometers away from the source, the fact of the matter is that today the cloud finds strong energy constraints because of the energy-hungry computing "factories." However, data centers are not the only computing resources that contribute to the energy inefficiency. Distributed computing devices and wireless communication layers are also responsible for this.

Energy efficiency in the cloud requires that the envisioned optimization techniques take into account the different layers of the computing paradigm, as well as the characteristics of the application and processed data. By providing horizontal and vertical optimization approaches, we can ensure that the total energy consumption reaches acceptable limits.

In this chapter, we reviewed several alternatives that, as opposed to traditional approaches, consider the total energy consumption of the whole set of resources that appear in cloud computing. These techniques provide a multilayer approach to tackle the problem of energy consumption and obtain bigger savings than any previous mechanism.

References

1. Dinh, H. T., Lee, C., Niyato, D., and Wang, P. 2011. A survey of mobile cloud computing: architecture, applications, and approaches. *Wireless Communications and Mobile Computing*, 13(18), 1587–1611.
2. Feng, W., and Scogland, T. 2009. The Green500 list: year one. In *IEEE International Symposium on Parallel and Distributed Processing (IPDPS)*, 1–7, IEEE Computer Society, Washington, D.C., USA.
3. Raghavendra, R., Ranganathan, P., Talwar, V., Wang, Z., and Zhu, X. 2008. No "power struggles": coordinated multi-level power management for the data center. *ACM SIGARCH Computer Architecture News*, 36(1), 48–59, 2008.
4. Kaplan, J. M., Forrest, W., and Kindler, N. 2008. *Revolutionizing Data Center Energy Efficiency*. Technical Report. New York: McKinsey & Company.
5. Bash, C., and Forman, G. 2007. Cool job allocation: measuring the power savings of placing jobs at cooling-efficient locations in the data center. *USENIX Annual Technical Conference*, 29, USENIX Association, Berkeley, CA, USA.
6. Koomey, J. 2011. *Growth in Data Center Electricity Use 2005 to 2010*. Oakland, CA: Analytics Press.
7. Ahmed, M. 2008. Google search finds seafaring solution. *The Times*, Sept. 15.
8. Vance, A. 2006. Microsoft's data center offensive sounds offensive. *The Register*, March 3.
9. Mullins, R. 2007. *HP Service Helps Keep Data Centers Cool*. Technical report. Boston: IDG News Service.
10. Dongarra, J. J., Luszczek, P., and Petitet, A. 2003. The LINPACK benchmark: past, present and future. *Concurrency and Computation: Practice and Experience* 15(9):803–820, John Wiley & Sons, Ltd., Geoffrey C. Fox and David W. Walker, eds.
11. Hamilton, J. 2009. Cooperative expendable micro-slice servers (CEMS): low cost, low power servers for internet-scale services. *Conference on Innovative Data Systems Research (CIDR'09)*, Asilomar, CA, Jan. 4–7.

12. Diaz, C. O., Guzek, M., Pecero, J. E., Bouvry, P., and Khan, S. U. 2011. Scalable and energy-efficient scheduling techniques for large-scale systems. *International Conference on Communications and Information Technology (ICCIT 2011)*, IEEE Computer Society, Washington, D.C., USA, 641–647.
13. Kliazovich, D., Bouvry, P., and Khan, S.-U. 2013. Dens: data center energy-efficient network-aware scheduling. *Cluster Computing*, 16, 65–75, Springer US.
14. Goiri, I., and Berral, J. L. 2012. Energy-efficient and multifaceted resource management for profit driven virtualized data centers. *FGCS* 28:718–731.
15. Quan, D. M., Mezza, F., Sannelli, D., and Giafreda, R. 2012. T-alloc: a practical energy efficient resource allocation algorithm for traditional data centers. *FGCS* 28:791–800.
16. Kusic, D., Kephart, J. O., Hanson, J. E., Kandasamy, N., and Jiang, G. 2009. Power and performance management of virtualized computing environments via lookahead control. *Cluster Computing* 12:1–15.
17. Wang, Y., and Wang, X. 2010. Power optimization with performance assurance for multi-tier applications in virtualized data centers. *Parallel Processing Workshops*, IEEE Computer Society, Washington, D.C., 512–519.
18. Banković, Z., et al. 2011. Bio-inspired enhancement of reputation systems for intelligent environments. *Information Sciences* 222:99–112.
19. Madhusudan, I., and Schmidt, R. 2009. Analytical modeling for thermodynamic characterization of data center cooling systems. *Journal of Electronic Packaging* 131:2.
20. Breen, T. J., et al. 2010. From chip to cooling tower data center modeling: Part I, Influence of server inlet temperature and temperature rise across cabinet. *Thermal and Thermomechanical Phenomena in Electronic Systems (ITherm), 12th IEEE Intersociety Conference*, 1–10.
21. Narendra, S. G., and Chandrakasan, A. P. 2006. *Leakage in Nanometer CMOS Technologies*. Heidelberg: Springer.
22. Rabaey, J. M. 2009. *Low Power Design Essentials*. New York: Springer.
23. Li, S., Abdelzaher, T., and Yuan, M. 2011. TAPA: temperature aware power allocation in data center with Map-Reduce. *Green Computing Conference and Workshops (IGCC)*, 1–8.
24. Zapater, M., et al. 2013. Leakage and temperature aware server control for improving energy efficiency in data centers. *Proceedings of the Conference on Design, Automation and Test in Europe*, 266–269.
25. Zapater, M., Ayala, J. L., and Moya, J. M. 2012. Leveraging heterogeneity for energy minimization in data centers. *Cluster, Cloud and Grid Computing (CCGrid)*, 752–757.
26. Li, T., and Lizy, K. J. 2003. Run-time modeling and estimation of operating system power consumption. *ACM SIGMETRICS*, 160–171.
27. Bircher, W. L., and Lizy, K. J. 2012. Complete system power estimation using processor performance events. *IEEE Transactions on Computers* 61(4):563–577.
28. Arroba, P., Zapater, M., Ayala, J. L., Moya, J. M., Olcoz, K., and Hermida, R. 2013. On the leakage-power modeling for optimal server operation. *Jornadas SARTECO*.
29. ASHRAE Technical Commitee. 2011. *Thermal Guidelines for Data Processing Environments*. Technical Report. Atlanta, GA: American Society of Heating, Refrigerating and Air-Conditioning Engineers.
30. Nathuji, R., Canturk I., and Gorbatov, E. 2007. Exploiting platform heterogeneity for power efficient data centers. *Autonomic Computing (ICAC'07)*, 5–5.

31. Zheng, X., and Yu, C. 2010. Markov model based power management in server clusters. *Green Computing and Communications (GreenCom)*, 96–102.
32. Bodenstein, C., Schryen G., and Neumann, D. 2011. Reducing datacenter energy usage through efficient job allocation. *European Council of International Schools (ECIS 2011)*, 108.
33. Bell, G. C. 2013. *Wireless Sensors Improve Data Center Efficiency.* DOE/Technical Case Study Bulletin, US Dept. of Energy, Washington, D.C., USA.
34. Abbasi, Z., et al. 2013. Evolutionary green computing solutions for distributed cyber physical systems. In *Evolutionary Based Solutions for Green Computing*, ed. Khan, S. U., Kołodziej, J., Li, J., and Zomaya, A. Y. New York: Springer, 1–28.
35. Henning, J. L. 2006. SPEC CPU2006 benchmark descriptions. *ACM SIGARCH Computer Architecture News* 34(4): 1–17.
36. Zapater, M., Sanchez, C., et al. 2012. Ubiquitous green computing techniques for high demand applications in smart environments. *Sensors* 12(8):10659–10677.
37. Garg, S. K., Gopalaiyengar, S. K., and Buyya, R. 2011. SLA-based resource provisioning for heterogeneous workloads in a virtualized cloud datacenter. *International Conference on Algorithms and Architectures for Parallel Processing*, 371–384.
38. MapReduce.org. 2011. What is MapReduce? http://www.mapreduce.org/what-is-mapreduce.php (accessed March 9, 2012).
39. Chen, Y., Keys, L., and Katz, R. H. 2009. *Towards Energy Efficient MapReduce.* Technical Report. Berkeley: EECS Department, University of California.
40. Gandhi, A., Harchol-Balter, M., Adan, I. 2010. Server farms with setup costs. *Performance Evaluation* 67(11):1123–1138.
41. Gandhi, A., Gupta, V., Harchol-Balter, M., and Kozuch, M. A. 2010. Optimality analysis of energy-performance trade-off for server farm management. *Performance Evaluation* 67(11):1155–1171.

Index